DRYWALL

W. Robert Harris

AMERICAN TECHNICAL PUBLISHERS, INC.
HOMEWOOD, ILLINOIS 60430

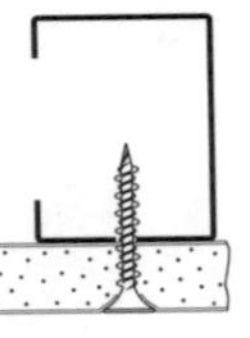

Acknowledgments

The author and publisher are grateful to the following companies for providing technical information and assistance.

- AMES Taping Tool Systems Inc.
- Hilti, Inc.
- Hyde Tools
- Klein Tools, Inc.
- Miller Electric Manufacturing Company
- National Gypsum Company
- Porter-Cable Corp.
- Red Devil, Inc.
- Ridge Tool Company
- Ruud Lighting, Inc.
- Spray Force Mfg., Inc.
- Stanley-Proto Industrial Tools
- United States Gypsum Company
- Wallboard Tool Co., Inc.

1 2 3 4 5 6 7 8 9 – 97 – 9 8 7 6 5 4 3 2 1

Printed in the United States of America

Harris, W. Robert (William Robert)
Drywall / W. Robert Harris.
p. cm.
Includes index.
ISBN 0-8269-0716-4 (soft)
1. Drywall. I. Title.
TH2239.H37 1997
690'.12--dc21

96-45664
CIP

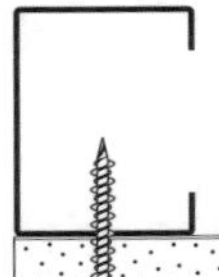

Contents

1 Trade Math 1
Trade Competency Test 25

2 Materials 29
Trade Competency Test 57

3 Tools and Equipment 61
Trade Competency Test 97

4 Drywall Handling and Installation 101
Trade Competency Test 119

5 Metal Framing Assemblies 121
Trade Competency Test 139

6 Sound- and Fire-Control Assemblies 143
Trade Competency Test 157

7 Special Installations and Materials 161
Trade Competency Test 175

8 Taping, Finishing, and Texturing 179
Trade Competency Test 191

Appendix 195

Glossary 209

Index 217

Introduction

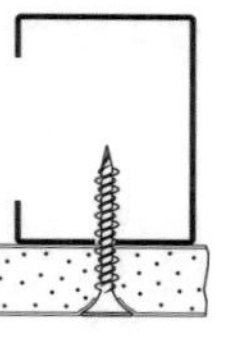

Drywall is a comprehensive text detailing the principles and practices of the drywall trade. Math concepts are presented with special emphasis on the math skills required to measure and cut drywall. Common materials and tools are described and shown in use.

Metal framing assemblies, sound- and fire-control assemblies, and special installations and materials are detailed. A chapter on taping, finishing, and texturing has been added. The Appendix contains many useful tables and the Glossary defines all technical terms used in the text. A comprehensive Index allows the user to quickly locate technical content.

The Publisher

1 Trade Math

Drywall workers use basic math concepts to estimate material and labor costs and to lay out and cut drywall. Whole numbers and common or decimal fractions in English or metric measurement systems are added, subtracted, multiplied, or divided to find solutions to problems. Sums, remainders, percentages, measurements, and area of plane figures are found by applying the proper math function(s).

WHOLE NUMBERS

Whole numbers are all numbers that have no fractional or decimal parts. See Figure 1-1. For example, numbers such as 1, 2, 19, 46, 67, 328, etc. are whole numbers. *Odd numbers* are any numbers that cannot be divided by 2 an exact number of times. For example, numbers such as 1, 3, 5, 57, 109, etc. are odd numbers. *Even numbers* are any numbers that can be divided by 2 an exact number of times. For example, numbers such as 2, 4, 6, 8, 10, 48, 432, etc. are even numbers.

Prime numbers are numbers that can be divided an exact number of times only by themselves and the number 1. For example, numbers such as 1, 2, 3, 5, 7, 11, 13, 17, 19, 23, etc. are prime numbers. Arabic and Roman numerals are the two common numeral systems used for calculations and notations.

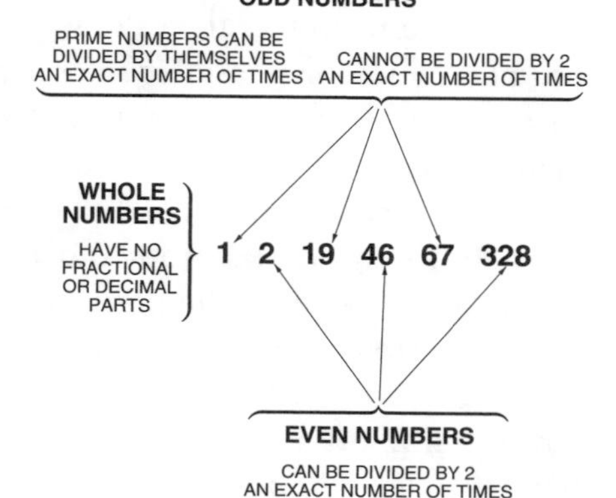

Figure 1-1. Whole numbers have no fractional or decimal parts.

A tape measure is used during drywall installation to determine the length to which the drywall sheets must be cut.

Arabic Numerals

Arabic numerals are numerals expressed by the ten digits 0, 1, 2, 3, 4, 5, 6, 7, 8, and 9. See Figure 1-2. These digits may be used alone or in combination to represent quantities indicating how much, how many, how far, how long, how hot, how expensive, etc. This is the numeral system most commonly used in the United States.

The Arabic numeral system is the most commonly used numeral system. Large Arabic numerals are made easier to read by the use of periods. A *period* is a group of three digits separated from other periods by a comma. The *units period* is the first period (000 through 999). The *thousands period* is the second period (1,000 through 999,999). The *millions period* is the third period (1,000,000 through 999,999,999).

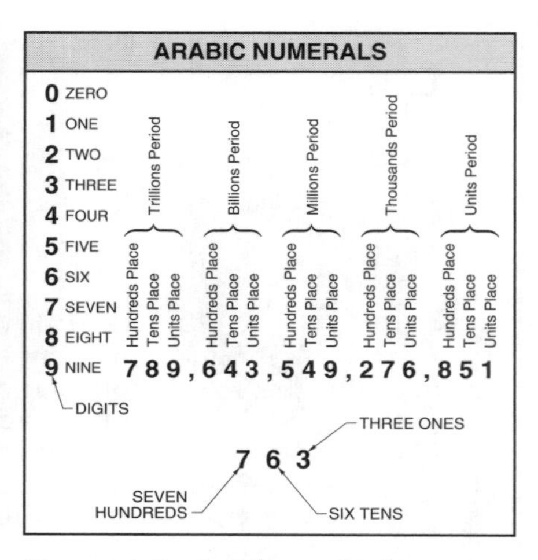

Figure 1-2. Arabic numerals are expressed by digits.

Roman Numerals

Roman numerals are numerals expressed by the letters I, V, X, L, C, D, and M. See Figure 1-3. While not commonly used in the trades, this numeral system is occasionally seen as chapter numbers in a book, on clock faces, and on public buildings such as libraries, museums, etc. In earlier times, various tradesworkers used Roman numerals to identify mating parts. For example, a carpenter could mark mating timbers by striking Roman numerals on them with a hammer and chisel. Roman numerals were easier to make with a hammer and chisel than Arabic numerals.

When a letter is followed by the same letter, or one lower in value, add the value of the letters. For example, XX = 20 and XV = 15. When a letter is followed by another letter greater in value, subtract the smaller letter. For example, IV = 4, IX = 9, and XC = 90.

ROMAN NUMERALS

Arabic	Roman	Arabic	Roman	Arabic	Roman
1	**I**	10	**X**	100	**C**
2	**II**	20	**XX**	200	**CC**
3	**III**	30	**XXX**	300	**CCC**
4	**IV**	40	**XL**	400	**CD**
5	**V**	50	**L**	500	**D**
6	**VI**	60	**LX**	600	**DC**
7	**VII**	70	**LXX**	700	**DCC**
8	**VIII**	80	**LXXX**	800	**DCCC**
9	**IX**	90	**XC**	900	**DCCCC or CM**
				1000	**M**
UNITS		TENS		HUNDREDS	

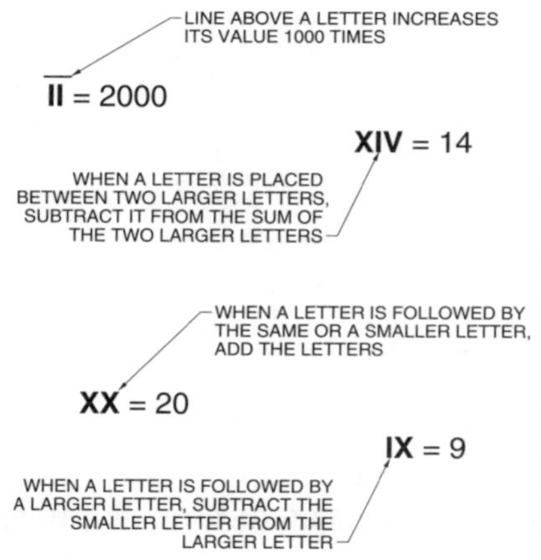

LINE ABOVE A LETTER INCREASES ITS VALUE 1000 TIMES

$\overline{\mathrm{II}}$ = 2000

XIV = 14

WHEN A LETTER IS PLACED BETWEEN TWO LARGER LETTERS, SUBTRACT IT FROM THE SUM OF THE TWO LARGER LETTERS

WHEN A LETTER IS FOLLOWED BY THE SAME OR A SMALLER LETTER, ADD THE LETTERS

XX = 20

IX = 9

WHEN A LETTER IS FOLLOWED BY A LARGER LETTER, SUBTRACT THE SMALLER LETTER FROM THE LARGER LETTER

Figure 1-3. Roman numerals are expressed by letters.

When a letter is placed between two letters of greater value, subtract the smaller letter from the sum of the other two. For example, XIV = 14. A superscript rule (line above) placed over a letter increases the value of the letter a thousand times. For example:

$\overline{\mathrm{V}}$ = 5000

$\overline{\mathrm{X}}$ = 10,000

Measurement Systems

Three common systems of measurement are the British (U.S.) system, decimal-inch system, and SI metric system (International System of Units). Arabic numerals are used with these three measurement systems. The British (U.S.) system is also known as the English system and is the system in primary use in the United States. This system uses the inch, foot, and pound units of measurement. The decimal-inch system is based on tenths and hundredths to simplify measurements. The decimal-inch system is used by surveyors, scientists, engineers, etc.

The metric system is the most common measurement system used in most of the world. See Figure 1-4. Prefixes are used in the metric system to represent multipliers. For example, the distance of 3000 meters is expressed as 3 kilometers. Metric measurements are converted to English measurements (and vice versa) by applying the appropriate conversion factors. See Appendix.

BASE UNITS		
Unit	**SI Symbol**	**Quantity**
Meter	m	Length
Gram	g	Mass
Second	s	Time
Ampere	A	Electric current

UNIT PREFIXES			
Prefix	**Unit**	**Symbol**	**Number**
Other larger multiples			
Mega	Million	M	1,000,000 = 10^6
Kilo	Thousand	k	1,000 = 10^3
Hecto	Hundred	h	100 = 10^2
Deka	Ten	d	10 = 10^1
			Unit 1 = 10^0
Deci	Tenth	d	0.1 = 10^{-1}
Centi	Hundredth	c	0.01 = 10^{-2}
Milli	Thousandth	m	0.001 = 10^{-3}
Micro	Millionth	μ	0.000001 = 10^{-6}
Other smaller multiples			

10	20	30	40	50	60	70	80	90	100

LENGTH

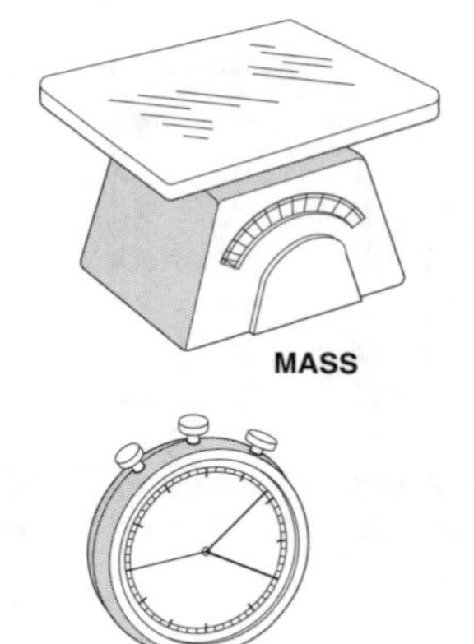

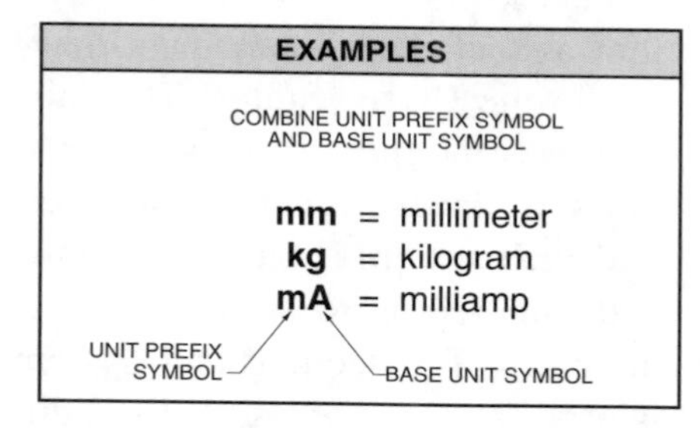

EXAMPLES
COMBINE UNIT PREFIX SYMBOL AND BASE UNIT SYMBOL
mm = millimeter
kg = kilogram
mA = milliamp
UNIT PREFIX SYMBOL
BASE UNIT SYMBOL

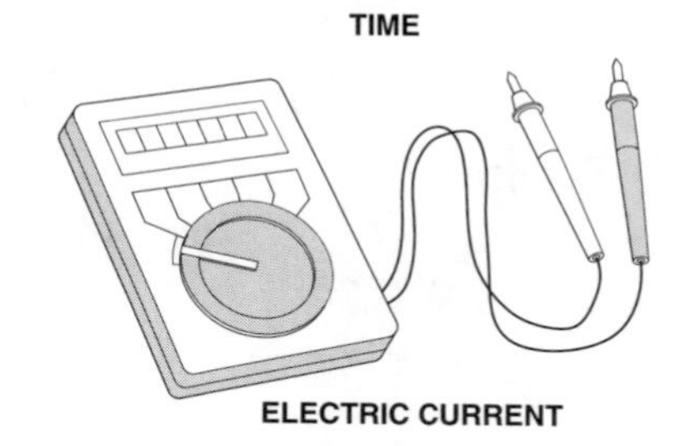

Figure 1-4. The metric system is the common measurement system used in most of the world.

Addition

Addition is the process of uniting two or more numbers to make one number. See Figure 1-5. It is the most common operation in mathematics. The sign + (plus) indicates addition and is used when numbers are added horizontally or when two numbers are added vertically. When more than two numbers are added vertically, the operation is apparent, and no sign is required. The *sum* is the result obtained from adding two or more numbers.

To add whole numbers vertically, place all numbers in aligned columns. The units must be in the ones (units) column, tens in the tens column, hundreds in the hundreds column, etc. Add the columns from top to bottom, beginning with the ones column. When the sum of the numbers in the ones column is 0 – 9, record the sum and add the tens column. When the sum of the numbers in the ones column is 10 or more, record the last digit and carry the remaining digit(s) to the tens column. Follow this same procedure for remaining columns.

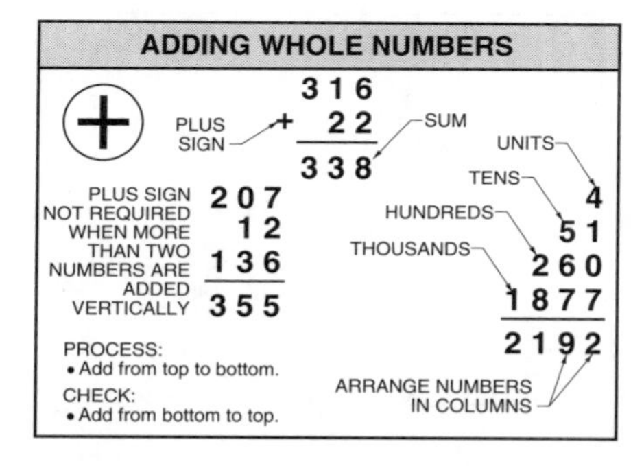

Figure 1-5. Addition is the process of uniting two or more numbers to make one number.

Adding whole numbers horizontally is more difficult than adding them vertically. For example, 25 + 120 + 37 + 3 = 185 shows whole numbers added horizontally. The horizontal alignment method is not as commonly used as the vertical alignment method because mistakes can occur more easily.

Vertically aligned addition problems are checked by adding the numbers from bottom to top. Horizontally aligned addition problems are checked by adding the numbers from right to left. The same sum occurs if both operations have been added correctly.

Subtraction

Subtraction is the process of taking one number away from another number. See Figure 1-6. It is the opposite of addition. The sign – (minus) indicates subtraction. The *minuend* is the number from which the subtraction is made. The *subtrahend* is the number which is subtracted. The *remainder* is the difference between the minuend and the subtrahend. Place the minuend above the subtrahend when vertically aligning numbers.

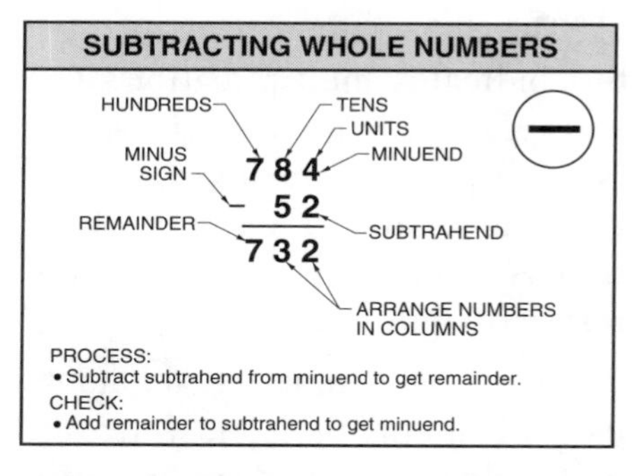

Figure 1-6. Subtraction is the process of taking one number away from another number.

As in addition, the first column of numbers represents ones, the second column represents tens, etc. When a subtrahend digit is larger than the corresponding minuend digit, borrow one unit (tens, hundreds, etc.) from the column immediately to the left and continue the operation. For example, when subtracting 8 from 24, borrow a 1 from the tens column, subtract 8 from 14, record the 6 in the units column, and record the remaining 1 in the tens column for a remainder of 16.

Multiplication

Multiplication is the process of adding one number as many times as there are units in the other number. See Figure 1-7. For example, 3 × 4 = 12 produces the same result as adding 4 + 4 + 4 = 12. The sign × (times or multiplied by) indicates multiplication. The *multiplicand* is the number which is multiplied. The *multiplier* is the number by which multiplication is done. The *product* is the result of multiplication.

The larger number is commonly used as the multiplicand when the units being multiplied are the same. For example, 8′ × 4′ = 32 sq ft. If a number to be multiplied represents a unit of measurement (inches, feet, pounds, etc.), identify the unit of measurement in the multiplicand, multiplier, and product. Numbers may be arranged vertically (preferred) or horizontally when multiplying. An effective method of checking the product is to reverse the multiplicand and the multiplier and perform the operation again. The same product occurs if both operations have been multiplied correctly.

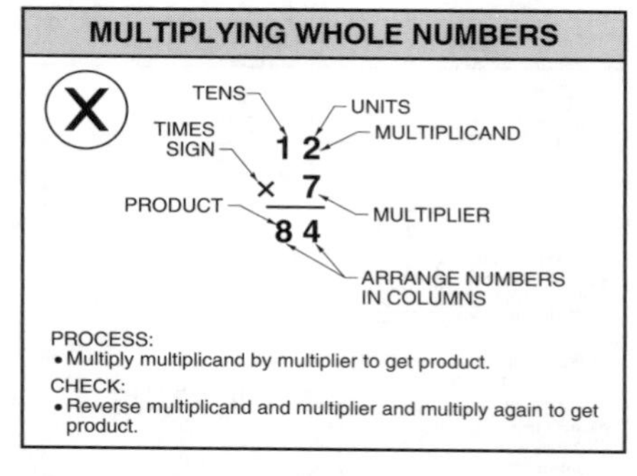

Figure 1-7. Multiplication is the process of adding one number as many times as there are units in the other number.

Zeros have no value, therefore any number multiplied by a zero equals zero. For example, 21 × 0 = 0. To multiply a multiplicand by 10, add one zero. For example, to multiply 74 by 10, add one zero to the 74 to get 740 (74 × 10 = 740). Add two zeros to multiply by 100, three zeros to multiply by 1000, etc.

Division

Division is the process of finding how many times one number contains the other number. It is the reverse of multiplication. See Figure 1-8. The sign ÷ (divided by) indicates division. The sign $\overline{)\quad}$ also indicates division. The *dividend* is the number to be divided. The *divisor* is the number by which division is done. The *quotient* is the result of division. The *remainder* is the part of the quotient left over whenever the quotient is not a whole number.

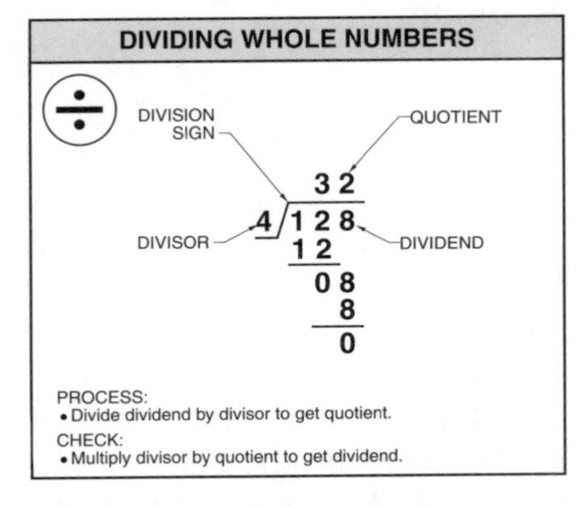

DIVIDING WHOLE NUMBERS
DIVISION SIGN
QUOTIENT
DIVISOR
DIVIDEND
32
4/128
12
08
8
0
PROCESS:
• Divide dividend by divisor to get quotient.
CHECK:
• Multiply divisor by quotient to get dividend.

Figure 1-8. Division is the process of finding how many times one number contains the other number.

To divide a number by 10, 100, etc. remove as many places from the right of the dividend as their are zeros in the divisor. For example, 500 ÷ 10 = 50. Notice that one zero was removed from the dividend (500) to yield the quotient of 50.

Any remainder is placed over the divisor and expressed as a fraction. For example, 27 ÷ 4 = $6\frac{3}{4}$. Notice that 4 goes into 27 six times with a remainder of 3. The 3 is placed over the 4 (divisor).

To check division, multiply the divisor by the quotient. For example, 48 ÷ 4 = 12. To check this problem, multiply 4 (divisor) by 12 (quotient). For example, 4 × 12 = 48.

COMMON FRACTIONS

A *fraction* is one part of a whole number. See Figure 1-9. The number 1 is the smallest whole number. Anything smaller than 1 is a fraction and can be divided into any number of fractional parts. Fractions are written above and below or on both sides of a fraction bar. Fraction bars may be horizontal or inclined.

The *denominator* is the part of a fraction that shows how many parts the whole number has been divided into. The denominator is the lower (or right-hand) number of a fraction. The *numerator* is part of a fraction that shows the number of parts in the fraction. The numerator is the upper (or left-hand) number. For example, the fraction

¾ shows that a whole number is divided into four equal parts (denominator), and three of these parts (numerator) are present.

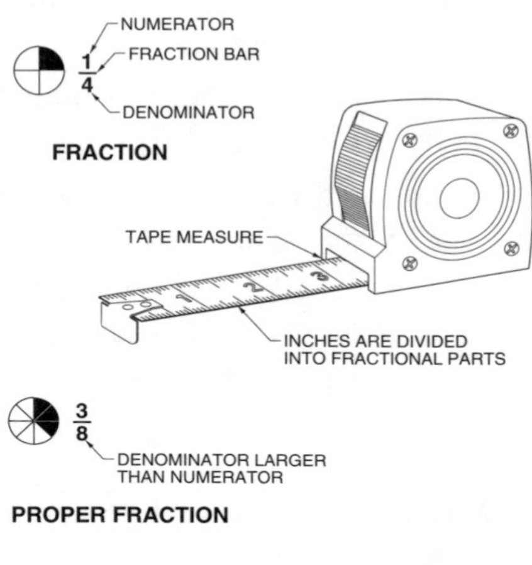

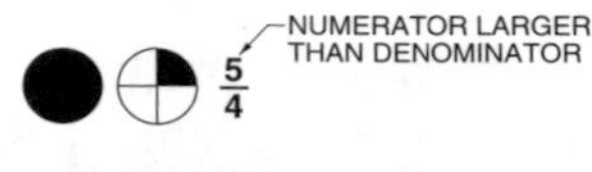

Figure 1-9. A fraction is one part of a whole number.

A *proper fraction* is a fraction that has a denominator larger than its numerator. An *improper fraction* is a fraction that has a numerator larger than its denominator. A *mixed number* is a combination of a whole number and a fraction. For example, ¾ is a proper fraction, 5⁄4 is an improper fraction, and 1¼ is a mixed number.

Any type of units can be divided into fractional parts. For example, inches are commonly divided into fractional parts of an inch based upon halves, fourths, eighths, sixteenths, thirty-seconds, and sixty-fourths. Fractional parts of an inch are always expressed in their lowest common denominator (LCD).

The lowest term is found by dividing the highest number that divides equally into the denominator and numerator. For example, the lowest term of the fraction 12⁄16 is ¾. This is obtained by dividing 4 into 12 and 4 into 16. Always reduce fractions to their lowest terms.

Addition

Fractions may be added horizontally or vertically. Horizontal placement is the most common, as identification of numerators and denominators is easier. Fractions which may be added include proper fractions, improper fractions, mixed numbers, and fractions with unlike denominators. There is a different rule for each of these four combinations. See Figure 1-10.

Adding Proper Fractions. Fractions having the same denominator are added by adding the numerators and placing them over the denominator. For example, in the problem ⅓ + ⅓ = ⅔, the numerators (1 + 1) are added to produce 2. The denominator (3) remains constant.

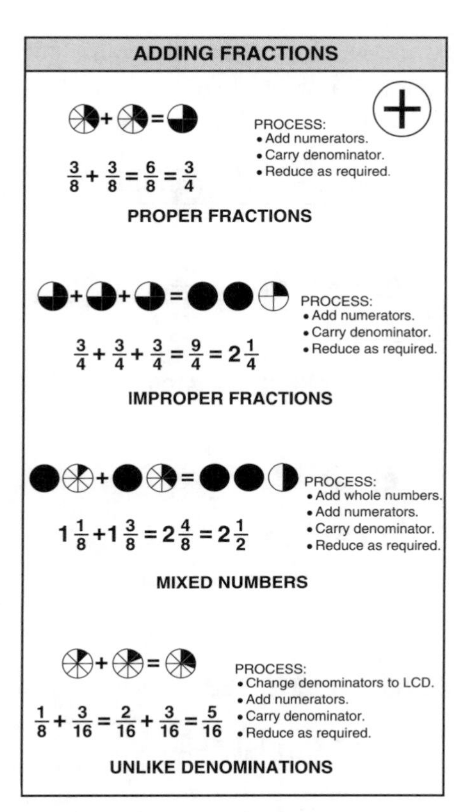
ADDING FRACTIONS
PROCESS:
• Add numerators.
• Carry denominator.
• Reduce as required.
$\frac{3}{8} + \frac{3}{8} = \frac{6}{8} = \frac{3}{4}$
PROPER FRACTIONS
PROCESS:
• Add numerators.
• Carry denominator.
• Reduce as required.
$\frac{3}{4} + \frac{3}{4} + \frac{3}{4} = \frac{9}{4} = 2\frac{1}{4}$
IMPROPER FRACTIONS
PROCESS:
• Add whole numbers.
• Add numerators.
• Carry denominator.
• Reduce as required.
$1\frac{1}{8} + 1\frac{3}{8} = 2\frac{4}{8} = 2\frac{1}{2}$
MIXED NUMBERS
PROCESS:
• Change denominators to LCD.
• Add numerators.
• Carry denominator.
• Reduce as required.
$\frac{1}{8} + \frac{3}{16} = \frac{2}{16} + \frac{3}{16} = \frac{5}{16}$
UNLIKE DENOMINATIONS

Figure 1-10. Fractions which may be added include proper or improper fractions, mixed numbers, and fractions with unlike denominators.

Adding Improper Fractions. Fractions which produce a sum in which the numerator is larger than the denominator (improper fractions) are changed to a mixed number by dividing the numerator by the denominator, recording the quotient obtained, and treating the remainder as a numerator placed over the original denominator. For example, in the problem $\frac{3}{8} + \frac{3}{8} + \frac{3}{8} = \frac{9}{8}$, the improper fraction ($\frac{9}{8}$) is changed to $1\frac{1}{8}$ by dividing 9 by 8. The fraction is then written $\frac{3}{8} + \frac{3}{8} + \frac{3}{8} = \frac{9}{8} = 1\frac{1}{8}$.

Adding Mixed Numbers. To add fractions containing mixed numbers, add the whole numbers, add the numerators, and carry the denominator. For example, in the problem $1\frac{1}{4} + 3\frac{1}{4} + 4\frac{1}{4} = 8\frac{3}{4}$, the whole numbers (1 + 3 + 4) are added to produce 8. The numerators (1 + 1 + 1) are added to produce 3, which is placed over the denominator 4.

Adding Fractions with Unlike Denominators. To add fractions in which the denominators are not the same, change the denominators to the LCD, add the numerators, and carry the denominator. If an improper fraction occurs, change the improper fraction to a mixed number. For example, to add $\frac{3}{8} + \frac{1}{2} + \frac{3}{4}$, change the denominators to 8 and multiply the number of times the original denominators go into 8 by the numerators to get $\frac{3}{8} + \frac{4}{8} + \frac{6}{8}$. Add the numerators 3 + 4 + 6 = 13 and place over the denominator to get $\frac{13}{8}$. Change the improper fraction $\frac{13}{8}$ by dividing 13 by 8. Thirteen can be divided by 8 one time with a remainder of 5, which is placed over the 8 to produce $1\frac{5}{8}$ ($\frac{3}{8} + \frac{1}{2} + \frac{3}{4} = \frac{3}{8} + \frac{4}{8} + \frac{6}{8} = \frac{13}{8} = 1\frac{5}{8}$).

Subtraction

Subtraction of fractions is similar to addition of fractions. Fractions may be subtracted horizontally or vertically. Horizontal placement is the most common, as identification of numerators and denominators is easier. All fractions must have a common denominator before one can be subtracted from another. Fractions which may be subtracted include fractions with like denominators, fractions with unlike denominators, and mixed numbers. There is a different rule for each of these three combinations. See Figure 1-11.

Subtracting Fractions with Like Denominators. To subtract fractions having the same denominators, subtract one numerator from the other numerator and place over the denominator. For example, to subtract $\frac{7}{16}$ from $\frac{11}{16}$, subtract the numerator 7 from the numerator 11 to get 4. Place the 4 over the denominator 16. Reduce $\frac{4}{16}$ by dividing by the largest number that goes into the numerator and denominator an even number of times. In this example, divide the numerator and denominator by 4 to get $\frac{1}{4}$ ($\frac{11}{16} - \frac{7}{16} = \frac{4}{16} = \frac{1}{4}$).

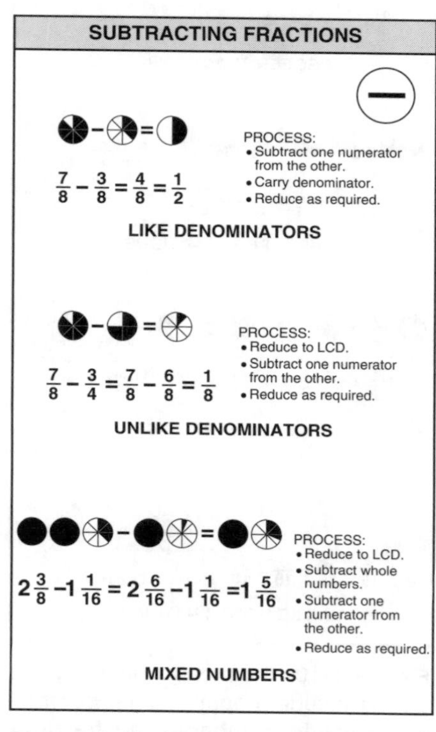

Figure 1-11. Fractions which may be subtracted include fractions with like or unlike denominators, and mixed numbers.

Subtracting Fractions with Unlike Denominators. To subtract fractions having unlike denominators, reduce the fractions to their LCD and subtract one numerator from the other. For example, to subtract $\frac{7}{16}$ from $\frac{3}{4}$, reduce $\frac{3}{4}$ to $\frac{12}{16}$ and subtract $\frac{7}{16}$ to get $\frac{5}{16}$ ($\frac{3}{4} - \frac{7}{16} = \frac{12}{16} - \frac{7}{16} = \frac{5}{16}$).

Subtracting Mixed Numbers. To subtract fractions having mixed numbers, follow the applicable procedure for denominators, subtract the numerators, subtract the whole numbers, and if necessary, reduce the fraction to its lowest common denominator. For example, to subtract $1\frac{1}{4}$ from $3\frac{1}{2}$, reduce the fractions to their LCD and subtract one numerator from another to get $\frac{2}{4} - \frac{1}{4} = \frac{1}{4}$. Subtract the whole numbers to get $3 - 1 = 2$. Add the whole number and the fraction to get $2 + \frac{1}{4} = 2\frac{1}{4}$.

Multiplication

Fractions may be multiplied horizontally or vertically. Horizontal placement of fractions is the most common, as identification of numerators and denominators is easier. Fractions which may be multiplied include two fractions, fractions and a whole number, a mixed number and a whole number, and two mixed numbers. There is a different rule for each of these four combinations. See Figure 1-12.

Multiplying Two Fractions. To multiply two fractions, multiply the numerator of one fraction by the numerator of the other fraction. Do the same with the denominators. Reduce the answer as required. For example, to multiply $\frac{3}{8}$ by $\frac{1}{8}$, multiply the numerators to get 3 ($3 \times 1 = 3$) and multiply the denominators to get 64 ($8 \times 8 = 64$). Thus, $\frac{3}{8} \times \frac{1}{8} = \frac{3}{64}$.

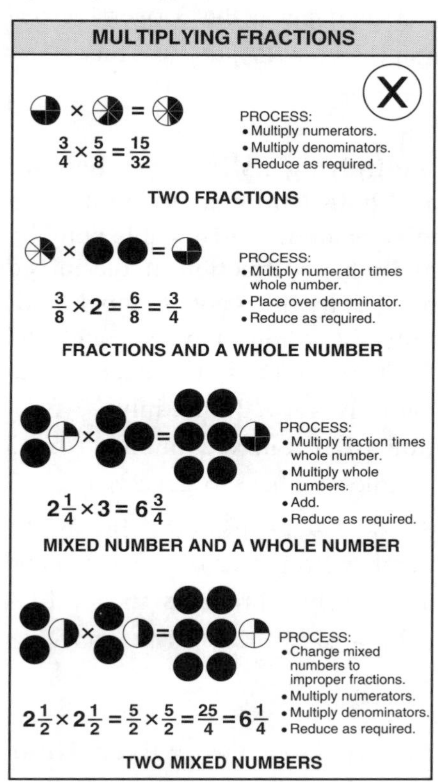

Figure 1-12. Fractions which may be multiplied include two fractions, fractions and a whole number, a mixed number and a whole number, and two mixed numbers.

Multiplying Fractions and a Whole Number. To multiply a fraction and a whole number, multiply the numerator of the fraction by the whole number and place over the denominator. Reduce the answer as required. For example, to multiply $\frac{1}{8} \times 3$, multiply the numerator 1×3 (whole number) to get 3 ($1 \times 3 = 3$) and place the 3 over the denominator 8 to get $\frac{3}{8}$. Thus, $\frac{1}{8} \times 3 = \frac{3}{8}$.

Multiplying a Mixed Number and a Whole Number. To multiply a mixed number and a whole number, multiply the fraction of the mixed number by the whole number, multiply the whole numbers, and add the two products. For example, to multiply $4\frac{7}{8} \times 3$, multiply $\frac{7}{8}$ (fraction of the mixed number) by 3 (whole number) to get $2\frac{5}{8}$ ($\frac{7}{8} \times 3 = \frac{21}{8} = 2\frac{5}{8}$). Multiply the whole numbers to get 12 ($4 \times 3 = 12$) and add the two products to get $14\frac{5}{8}$ ($2\frac{5}{8} + 12 = 14\frac{5}{8}$). Thus, $4\frac{7}{8} \times 3 = 14\frac{5}{8}$.

Multiplying Two Mixed Numbers. To multiply two mixed numbers, change both mixed numbers to improper fractions and multiply. For example, to multiply $3\frac{1}{4}$ by $4\frac{1}{2}$, change the mixed number $3\frac{1}{4}$ to the improper fraction $\frac{13}{4}$ by multiplying the whole number 3 by the denominator 4 and adding the 1 ($3 \times 4 = 12 + 1 = \frac{13}{4}$). Change the mixed number $4\frac{1}{2}$ to the improper fraction $\frac{9}{2}$ by multiplying the whole number 4 by the denominator 2 and adding the 1 ($4 \times 2 = 8 + 1 = \frac{9}{2}$). Multiply the improper fractions to get $12\frac{1}{8}$ ($\frac{13}{4} \times \frac{9}{2} = \frac{117}{8} = 14\frac{5}{8}$). Thus, $3\frac{1}{4} \times 4\frac{1}{2} = 14\frac{5}{8}$.

Division

Fractions are divided horizontally. Fractions which may be divided include a fraction by a whole number, a mixed number by a whole number, two fractions, a whole number by a fraction, and two mixed numbers. There is a different rule for each of these five combinations. See Figure 1-13.

Dividing a Fraction by a Whole Number. To divide a fraction by a whole number, multiply the denominator of the fraction by the whole number. For example, to divide $\frac{3}{8}$ by 4, multiply the denominator 8 by the whole number 4 to get 32 ($8 \times 4 = 32$). Place the numerator 3 over the 32 to get $\frac{3}{32}$. Thus, $\frac{3}{8} \div 4 = \frac{3}{32}$.

Dividing a Mixed Number by a Whole Number. To divide a mixed number by a whole number, change the mixed number to an improper fraction and multiply the denominator of the improper fraction by the whole number. For example, to divide $2\frac{7}{8}$ by 3, change the mixed number $2\frac{7}{8}$ to $\frac{23}{8}$. Multiply the denominator of the improper fraction $\frac{23}{8}$ by the whole number 3 to get $\frac{23}{24}$. Thus, $2\frac{7}{8} \div 3 = \frac{23}{24}$.

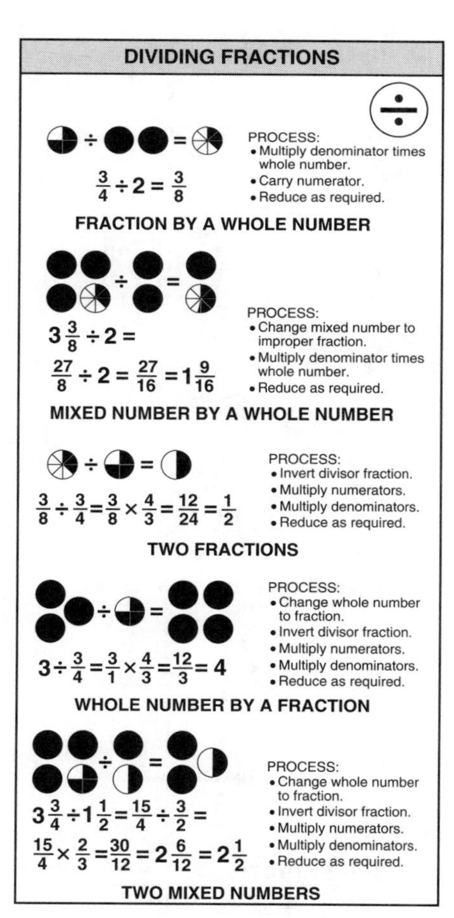

Figure 1-13. Fractions which may be divided include a fraction by a whole number, a mixed number by a whole number, two fractions, a whole number by a fraction, and two mixed numbers.

Dividing Two Fractions. To divide two fractions, invert the divisor fraction and multiply the numerator by the numerator and the denominator by the denominator. For example, to divide $\frac{3}{8}$ by $\frac{1}{4}$, invert the divisor fraction $\frac{1}{4}$ and multiply by $\frac{3}{8}$ to get $1\frac{1}{2}$ ($\frac{3}{8} \times \frac{4}{1} = \frac{12}{8} = 1\frac{4}{8} = 1\frac{1}{2}$). Thus, $\frac{3}{8} \div \frac{1}{4} = 1\frac{1}{2}$.

Dividing a Whole Number by a Fraction. To divide a whole number by a fraction, change the whole number into fraction form, invert the divisor fraction, and multiply the numerator by the numerator and the denominator by the denominator. For example, to divide 12 by $\frac{3}{4}$, change the whole number 12 to fraction form $\frac{12}{1}$. Invert the divisor fraction $\frac{3}{4}$ and multiply to get 16 ($\frac{12}{1} \times \frac{4}{3} = \frac{48}{3} = 16$). Thus, $12 \div \frac{3}{4} = 16$.

Dividing Two Mixed Numbers. To divide two mixed numbers, change both mixed numbers to improper fractions, invert the divisor fraction, and multiply the numerator by the numerator and the denominator by the denominator. For example, to divide $12\frac{1}{2}$ by $3\frac{1}{8}$, change the mixed number $12\frac{1}{2}$ to $\frac{25}{2}$ and the mixed number $3\frac{1}{8}$ to $\frac{25}{8}$. Invert the divisor fraction $\frac{25}{8}$ and multiply to get 4 ($\frac{25}{2} \times \frac{8}{25} = \frac{200}{50} = 4$). Thus, $12\frac{1}{2} \div 3\frac{1}{8} = 4$.

DECIMALS

A *decimal* is a fraction with a denominator of 10, 100, 1000, etc. See Figure 1-14. The number 1 is the smallest whole number. Anything smaller than 1 is a decimal and can be divided into any number of decimal parts. For example, the decimal .75 shows that the whole number 1 is divided into 100 equal parts, and 75 of these parts are present. Any fraction with 10, 100, 1000, or other multiple of ten for the denominator, may be written as a decimal. For example, the fraction $\frac{1}{10}$ is .1 in decimals, $\frac{1}{100}$ is .01, and $\frac{1}{1000}$ is .001.

A *decimal point* is the period in a decimal number. Shop workers and others who use decimals in their work often say "point" at the decimal point. For example, to denote 3.22, the worker says "three point twenty-two." This denotes $3\frac{22}{100}$. Others may say *and* at the decimal point. For example, to denote 4.37, they may say "four and thirty-seven hundredths." Either of the methods of expressing decimals is acceptable.

The United States monetary system is based on decimals. The dollar ($1.00) is valued at 100 cents. Each cent is $\frac{1}{100}$ of a dollar or $.01. Each nickel is $\frac{5}{100}$ of a dollar or $.05. Each dime is $\frac{10}{100}$ of a dollar or $.10. Each quarter is $\frac{25}{100}$ of a dollar or $.25. Each half-dollar is $\frac{50}{100}$ of a dollar or $.50.

DECIMALS			
Currency	**Value**	**Decimal**	**Fraction**
DOLLAR BILL	$1.00	1.00	$\frac{100}{100}$
HALF-DOLLAR	50¢	.50	$\frac{50}{100}$
QUARTER	25¢	.25	$\frac{25}{100}$
DIME	10¢	.10	$\frac{10}{100}$
NICKEL	5¢	.05	$\frac{5}{100}$
PENNY	1¢	.01	$\frac{1}{100}$

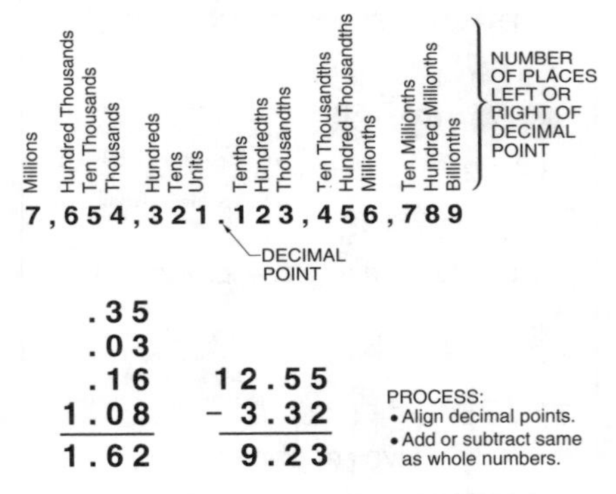

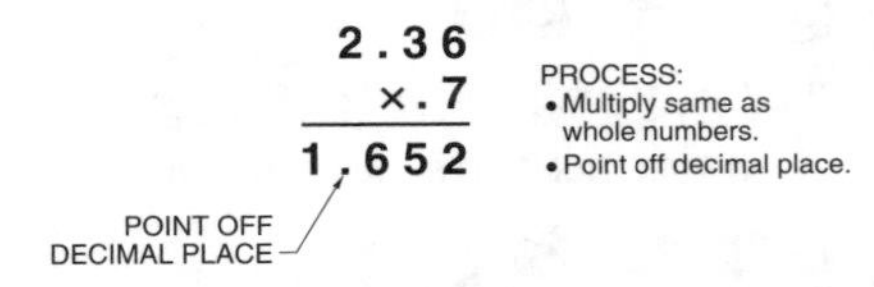

3.
3.25./9.75.
9 75

PROCESS:
• Divide same as whole numbers.
• Point off decimal place.

DIVIDING DECIMALS

Figure 1-14. A decimal is a fraction with a denominator of 10, 100, 1000, 10,000, etc.

More places written in a decimal indicate a higher degree of accuracy. For example, while .2 and .20 represent the same value, .20 is measured in hundredths and .2 is measured in tenths.

Adding or Subtracting Decimals

To add or subtract decimals, align the numbers vertically on the decimal points. Thus are units added or subtracted to units, tenths to tenths, hundredths to hundredths, etc. Add or subtract as in whole numbers and place the decimal point of the sum or remainder directly below the other decimal points. For example, to add 27.08 and 9.127, align the numbers vertically on the decimal points and add to get 36.207.

Multiplying Decimals

To multiply decimals, multiply as in whole numbers. Then, beginning at the right of the product, point off to the left the same number of decimal places in the quantities multiplied.

Prefix zeros when necessary. For example, to multiply 20.45 by 3.15, align the numbers vertically on the decimal points, multiply as in whole numbers, and point off four places from the right of the decimal point to get 64.4175.

Dividing Decimals

To divide decimals, divide as though dividend and divisor are whole numbers. Then point off from right to left as many decimal places as the difference between the number of decimal places in the dividend and divisor.

If the dividend has less decimal places than the divisor, add zeros to the dividend. There must be at least as many decimal places in the dividend as in the divisor. For example, to divide 2.5 into 16.75, divide as in whole numbers and point off one decimal place from right to left to get 6.7.

Changing Decimals to Common Fractions

To change a decimal to a common fraction, use the figures in the quantity as a numerator. For the denominator, place the figure 1 followed by as many zeros as there are figures to the right of the decimal point in the quantity. For example, to change the decimal .4 to a common fraction, place the 4 as a numerator and place a 1 followed by a zero as the denominator to get $^{4}/_{10}$.

To change the decimal .47 to a common fraction, place the 47 as a numerator and place a 1 followed by two zeros as the denominator to get $^{47}/_{100}$. To change the decimal

.479 to a common fraction, place the 479 as a numerator and place a 1 followed by three zeros as the denominator to get $^{479}/_{1000}$.

PLANE FIGURES

A *plane figure* is a flat figure. It has no depth. Plane figures are the basis for sketching and drawing. All plane figures are composed of lines drawn at various angles or with arcs. Plane figures include circles, triangles, quadrilaterals, and polygons. A *regular plane figure* has equal angles and equal sides. An *irregular plane figure* does not have equal angles and equal sides.

Lines

A *straight line* is the shortest distance between two points. See Figure 1-15. It is commonly referred to as a line. A *horizontal line* is parallel to the horizon. It is a level line. A *vertical line* is perpendicular to the horizon. Vertical, perpendicular, and plumb are at right angles to a baseline. *Vertical* is a line in a straight upward position. *Perpendicular* stresses the straightness of a line making a right angle with another (not necessarily horizontal) line. The symbol for perpendicular is ⊥. *Plumb* is an exact verticality (determined by a plumb bob and line) with Earth's gravity.

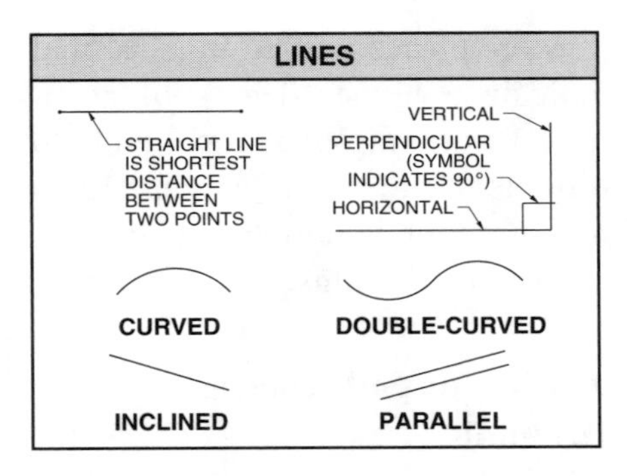

Figure 1-15. A straight line is the shortest distance between two points.

Lines may be drawn in any position. An *inclined (slanted) line* is neither horizontal nor vertical. *Parallel lines* remain the same distance apart. The symbol for parallel lines is ||.

Angles. An *angle* is the intersection of two lines. The symbol for angle is ∠. Angles are measured in degrees, minutes, and seconds. The symbol for degrees is °. The symbol for minutes is ′. The symbol for seconds is ″. There are 360° in a circle. There are 60 minutes in one degree and 60 seconds in one minute. For example, an angle might contain 112°-30′-12″.

A *straight angle* contains 180°. A *right angle* contains 90°. An *acute angle* contains less than 90°. An *obtuse angle* contains more than 90°. *Complementary angles* equal 90°. *Supplementary angles* equal 180°. See Figure 1-16.

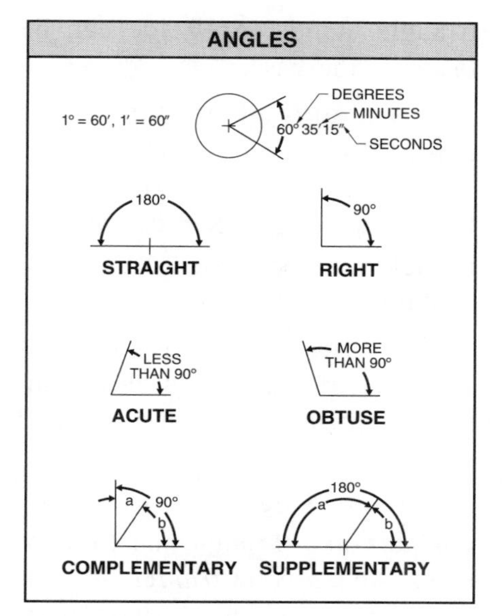

Figure 1-16. An angle is the intersection of two lines.

Circles

A *circle* is a plane figure generated around a centerpoint. See Figure 1-17. All circles contain 360°. The *diameter* is the distance from circumference (outside) to circumference through the centerpoint. The *circumference* is 3.1416 times the diameter of a circle. The *radius* is one-half the length of the diameter.

A *chord* is a line from circumference to circumference not through the centerpoint. An *arc* is a portion of the circumference. A *quadrant* is one-fourth of a circle. Quadrants have a right angle. A *sector* is a pie-shaped piece of a circle. A *segment* is the portion of a circle set off by a chord. A *semicircle* is one-half of a circle. Semicircles always contain 180°. *Concentric circles* have different diameters and the same centerpoint. *Eccentric circles* have different diameters and different centerpoints.

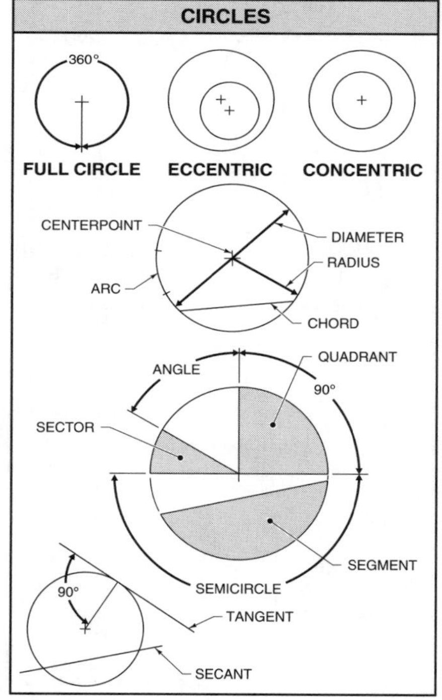

Figure 1-17. A circle is a plane figure generated around a centerpoint.

A *tangent* is a straight line touching the circumference at only one point. It is 90° to the radius. A *secant* is a straight line touching the circumference at two points.

Triangles

A *triangle* is a three-sided polygon with three interior angles. The sum of the three angles of a triangle is always 180°. The sign (Δ) indicates a triangle. See Figure 1-18.

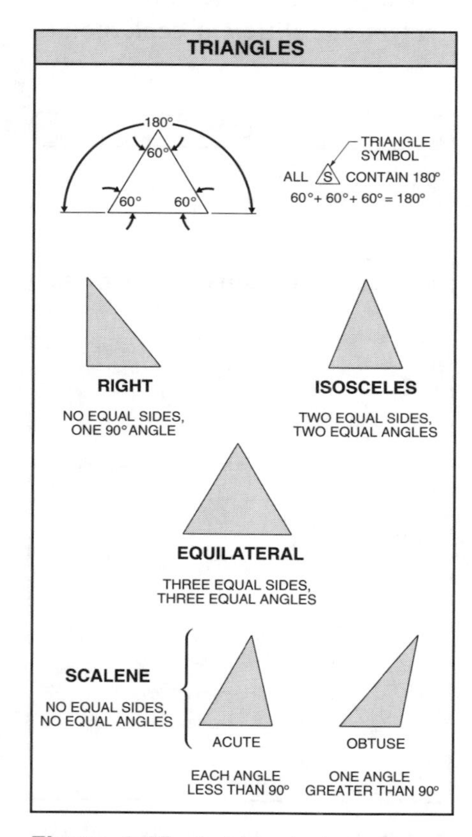

Figure 1-18. A triangle is a three-sided plane figure.

The *altitude* of a triangle is the perpendicular dimension from the vertex to the base. The *base* of a triangle is the side upon which the triangle stands. Any side can be taken as the base.

The angles of a triangle are named by uppercase letters. The sides of a triangle are named by lowercase letters. For example, a triangle may be named ΔABC and contain sides d, e, and f.

The different kinds of triangles are right triangles, isosceles triangles, equilateral triangles, and scalene triangles.

A *right triangle* is a triangle that contains one 90° angle and no equal sides. An *isosceles triangle* is a triangle that contains two equal angles and two equal sides.

An *equilateral triangle* is a triangle that has three equal angles and three equal sides. Each angle of an equilateral triangle is 60°.

A *scalene triangle* is a triangle that has no equal angles or equal sides. A scalene triangle may be acute or obtuse. An *acute triangle* is a scalene triangle with each angle less than 90°. An *obtuse triangle* is a scalene triangle with one angle greater than 90°.

Quadrilaterals

A *quadrilateral* is a four-sided polygon with four interior angles. The sum of the four angles of a quadrilateral is always 360°. The kinds of quadrilaterals are squares, rectangles, rhombuses, rhomboids, trapezoids, and trapeziums. See Figure 1-19.

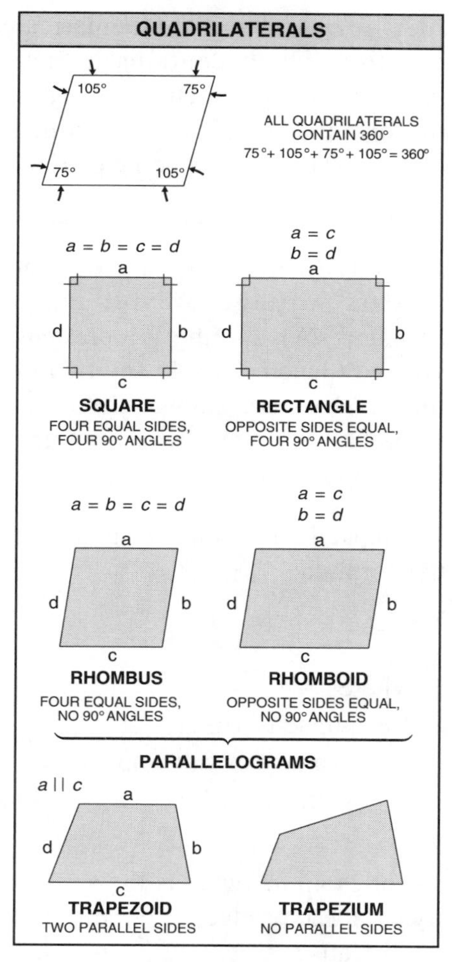

Figure 1-19. A quadrilateral is a four-sided plane figure.

A *square* is a quadrilateral with all sides equal and four 90° angles. A *rectangle* is a quadrilateral with opposite sides equal and four 90° angles. A *rhombus* is a quadrilateral with all sides equal and no 90° angles. A *rhomboid* is a quadrilateral with opposite sides equal and no 90° angles.

The square, rectangle, rhombus, and rhomboid are parallelograms. A *parallelogram* is a four-sided plane figure with opposite sides parallel and equal.

A *trapezoid* is a quadrilateral with two sides parallel. A *trapezium* is a quadrilateral with no sides parallel. Trapezoids and trapeziums are not parallelograms because all opposite sides are not parallel.

Polygons

A *polygon* is a many-sided plane figure. See Figure 1-20. All polygons are bounded by straight lines. A *regular polygon* has equal sides and equal angles. An *irregular polygon* has unequal sides and unequal angles. Polygons are named according to their number of sides. For example, a triangle has three sides; a quadrilateral has four sides; a pentagon has five sides; a hexagon has six sides; a heptagon has seven sides; an octagon has eight sides; etc.

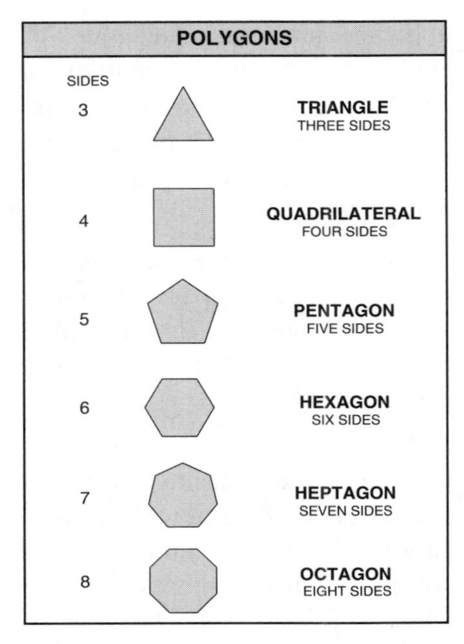

Figure 1-20. A polygon is a many-sided plane figure.

BASIC MATH FORMULAS

An *equation* is a means of showing that two numbers, or two groups of numbers, are equal to the same amount. For example, a baker has two pies of the same size. Pie A is cut into six equal pieces, and Pie B is cut into eight equal pieces. One customer buys three pieces of Pie A, and another customer buys four pieces of Pie B. The customers have bought the same amount of pie because $3/6 = 1/2$ and $4/8 = 1/2$. All equations must balance, and as $3/6$ and $4/8$ each equal $1/2$, $3/6 = 4/8$ is an equation.

A *formula* is a mathematical equation which contains a fact, rule, or principle. Letters are used in formulas to represent values (amount). In the common electrical formula, $I = \frac{VA}{V}$, the I represents ampacity (denoted A), the VA represents wattage or volt amps (denoted VA), and the V represents voltage (denoted V). If any two of these values are known, the other value can be found by rearranging the formula. See Figure 1-21.

Ampacity is found by applying the formula:

$$I = \frac{VA}{V}$$

where

I = ampacity (in A)

VA = wattage or volt amps (in VA)

V = voltage (in V)

For example, what is the ampacity of a 120 V electric circuit with 2400 watts?

$$I = \frac{VA}{V}$$

$$I = \frac{2400}{120}$$

$$I = \mathbf{20\ A}$$

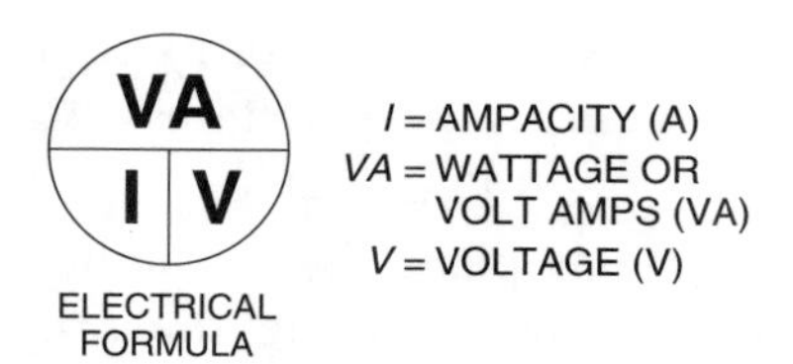

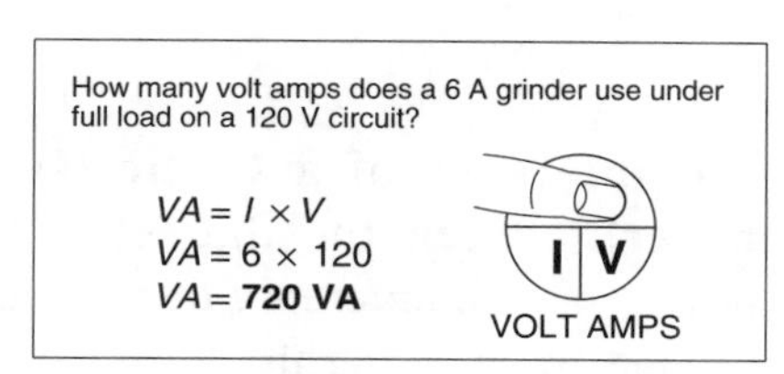

How many volt amps does a 6 A grinder use under full load on a 120 V circuit?
VA = I × V
VA = 6 × 120
VA = 720 VA
I
V
VOLT AMPS

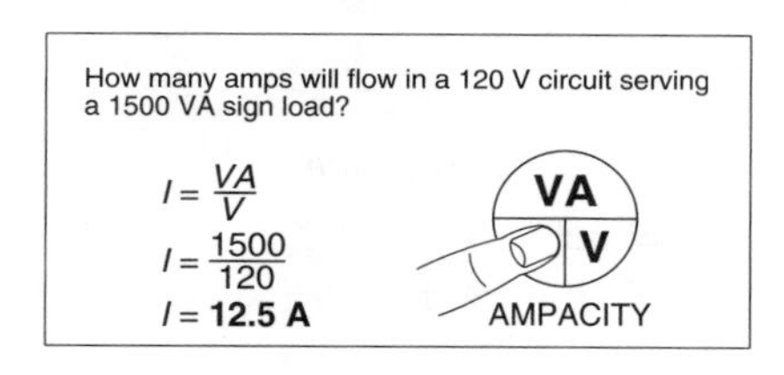

How many amps will flow in a 120 V circuit serving a 1500 VA sign load?
I = VA/V
I = 1500/120
I = 12.5 A
VA
V
AMPACITY

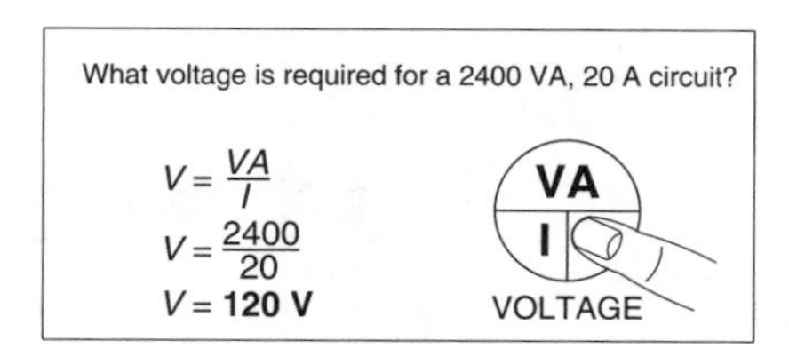

What voltage is required for a 2400 VA, 20 A circuit?
V = VA/I
V = 2400/20
V = 120 V
VA
I
VOLTAGE

Figure 1-21. When two values of a formula are known, the third value can be found.

Voltage is found by applying the formula:

$$V = \frac{VA}{I}$$

For example, what is the voltage of a 15 A circuit with 1800 VA?

$$V = \frac{VA}{I}$$

$$V = \frac{1800}{15}$$

$$V = \mathbf{120\ V}$$

Wattage is found by applying the formula:

$$VA = I \times V$$

For example, what is the volt amps of a 120 V, 30 A circuit?

$$VA = I \times V$$

$$VA = 30 \times 120$$

$$VA = \mathbf{3600\ VA}$$

Greek Letters

The letters of the Greek alphabet are frequently used in math formulas. For example, the Greek letter Pi represents 3.1416, the circumference of a circle. Pi is written as π. Uppercase and lowercase letters are used in the Greek alphabet. See Figure 1-22.

GREEK LETTERS					
Αα	ALPHA	Ιι	IOTA	Ρρ	RHO
Ββ	BETA	Κκ	KAPPA	Σσ	SIGMA
Γγ	GAMMA	Λλ	LAMBDA	Ττ	TAU
Δδ	DELTA	Μμ	ZETA	Υυ	UPSILON
Εε	EPSILON	Νν	NU	Φφ	PHI
Ζζ	ZETA	Ξξ	XI	Χχ	CHI
Ηη	ETA	Οο	OMICRON	Ψψ	PSI
Θθ	THETA	Ππ	PI	Ωω	OMEGA

Figure 1-22. The letters of the Greek alphabet are frequently used in math formulas.

Common Formulas

In the trades, common formulas are used when laying out jobs. See Figure 1-23. For example, a circular area may require a drywall finish. By applying the area formula for circles, the area of the circle can be determined.

Area. *Area* is the number of unit squares equal to the surface of an object. For example, a standard size piece of drywall contains 32 sq ft (4 × 8 = 32 sq ft). Area is expressed in square inches, square feet, and other units of measure. A *square inch* measures 1″ × 1″ or its equivalent. A *square foot* contains 144 sq in. (12″ × 12″ = 144 sq in.). The area of any plane figure can be determined by applying the proper formula.

Circumference of a Circle (Diameter). When the diameter is known, the circumference of a circle is found by applying the formula:

$C = \pi D$
where
C = circumference
π = 3.1416
D = diameter

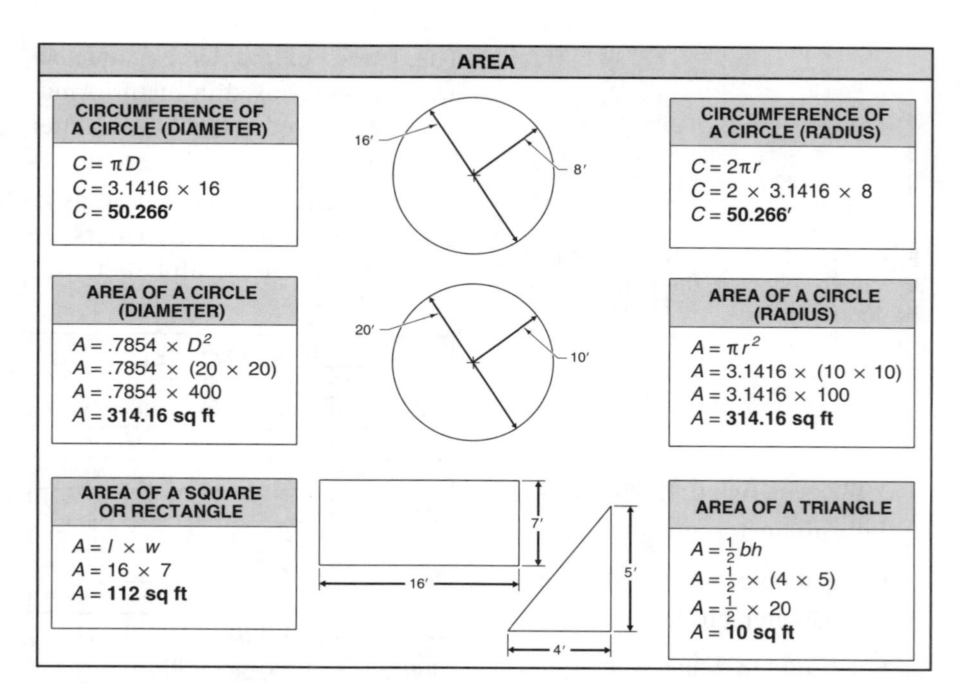

Figure 1-23. Plane formulas are used to find circumference and area.

For example, what is the circumference of a 20′ diameter circle?

$C = \pi D$
$C = 3.1416 \times 20$
$C = \mathbf{62.832'}$

Circumference of a Circle (Radius). When the radius is known, the circumference of a circle is found by applying the formula:

$C = 2\pi r$
where
C = circumference
2 = constant
$\pi = 3.1416$
r = radius

For example, what is the circumference of a 10′ radius circle?

$C = 2\pi r$
$C = 2 \times 3.1416 \times 10$
$C = \mathbf{62.832'}$

Area of a Circle (Diameter). When the diameter is known, the area of a circle is found by applying the formula:

$A = .7854 \times D^2$
where
A = area
$.7854$ = constant
D^2 = diameter squared

For example, what is the area of a 28′ diameter circle?

$A = .7854 \times D^2$
$A = .7854 \times (28 \times 28)$
$A = .7854 \times 784$
$A =$ **615.754 sq ft**

Area of a Circle (Radius). When the radius is known, the area of a circle is found by applying the formula:

$A = \pi r^2$
where
A = area
$\pi = 3.1416$
r^2 = radius squared

For example, what is the area of a 14′ radius circle?

$A = \pi r^2$
$A = 3.1416 \times (14 \times 14)$
$A = 3.1416 \times 196$
$A =$ **615.754 sq ft**

Area of a Square or Rectangle. The area of a square or the area of a rectangle is found by applying the formula:

$A = l \times w$
where
A = area
l = length
w = width

For example, what is the area of a 22′-0″ × 16′-0″ storage room?

$A = l \times w$
$A = 22 \times 16$
$A =$ **352 sq ft**

Area of a Triangle. The area of a triangle is found by applying the formula:

$A = \frac{1}{2}bh$

where

A = area

$\frac{1}{2}$ = constant

b = base

h = height

For example, what is the area of a triangle with a 10′-0″ base and a 12′-0″ height?

$A = \frac{1}{2}bh$

$A = \frac{1}{2} \times (10 \times 12)$

$A = \frac{1}{2} \times 120$

$A = \mathbf{60\ sq\ ft}$

Pythagorean Theorem. The *Pythagorean Theorem* states that the square of the hypotenuse of a right triangle is equal to the sum of the squares of the other two sides. The *hypotenuse* is the side of a right triangle opposite the right angle. Because a right triangle can have a 3-4-5 relationship, it is often used in laying out right angles and checking corners for squareness. See Figure 1-24.

The length of the hypotenuse of a right triangle is found by applying the formula:

$c = \sqrt{a^2 + b^2}$

where

c = length of hypotenuse

a^2 = length of one side squared

b^2 = length of other side squared

For example, what is the length of the hypotenuse of a triangle having sides of 3′ and 4′?

$c =$

$c = \sqrt{(3 \times 3) + (4 \times 4)}$

$c = \sqrt{9 + 16}$

$c = \sqrt{25}$

$c = \mathbf{5'}$

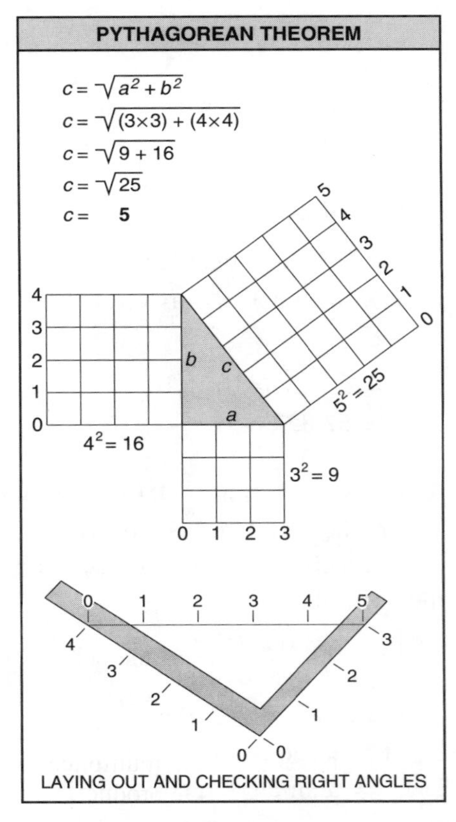

PYTHAGOREAN THEOREM
c = √(a² + b²)
c = √((3×3) + (4×4))
c = √(9 + 16)
c = √25
c = 5
4² = 16
3² = 9
5² = 25
b
c
a
LAYING OUT AND CHECKING RIGHT ANGLES

Figure 1-24. The square of the hypotenuse of a right triangle is equal to the sum of the squares of the other two sides of the triangle.

Trade Math

Trade Competency Test

Name   Date

Trade Math

_________ **1.** _________ numbers are all numbers that have no fractional or decimal parts.

_________ **2.** _________ numbers are numbers that can be divided an exact number of times only by themselves and the number 1.

_________ **3.** _________ numerals are numerals expressed by the ten digits 0, 1, 2, 3, 4, 5, 6, 7, 8, and 9.

_________ **4.** A(n) _________ is a group of three digits separated from other periods by a comma.

_________ **5.** The _________ is the number from which the subtraction is made.

_________ **6.** The _________ is the number which is subtracted.

A. minuend
B. subtrahend
C. remainder
D. neither A, B, nor C

_________ **7.** The _________ is the difference between the minuend and the subtrahend.

_________ **8.** The _________ is the number which is multiplied.

A. multiplicand
B. product
C. multiplier
D. neither A, B, nor C

T F **9.** The multiplier is the number by which multiplication is done.

_________ **10.** _________ numerals are numerals expressed by the letters I, V, X, L, C, D, and M.

_________ 11. The _________ is the result of multiplication.

A. multiplicand
B. product
C. multiplier
D. neither A, B, nor C

_________ 12. The _________ is the number to be divided.

_________ 13. The _________ is the number by which division is done.

A. dividend
B. quotient
C. divisor
D. neither A, B, nor C

T F 14. The dividend is the result of division.

_________ 15. The _________ is the part of the quotient left over whenever the quotient is not a whole number.

_________ 16. A(n) _________ is one part of a whole number.

_________ 17. The _________ is the part of a fraction that shows how many parts the whole number has been divided into.

A. numerator
B. quotient
C. denominator
D. neither A, B, nor C

T F 18. The denominator is the part of a fraction that shows the number of parts in the fraction.

_________ 19. A(n) _________ fraction is a fraction that has a denominator larger than its numerator.

_________ 20. A(n) _________ fraction is a fraction that has a numerator larger than its denominator.

_________ 21. A(n) _________ line is the shortest distance between two points.

_________ 22. A(n) _________ is the intersection of two lines.

_________ 23. _________ lines remain the same distance apart.

_________ 24. A(n) _________ angle contains 90°.

A. right
B. obtuse
C. straight
D. complimentary

_________ 25. A(n) _________ angle contains less than 90°.

T F 26. An obtuse angle contains more than 90°.

__________ **27.** _________ angles equal 90°.

A. Right
B. Obtuse
C. Straight
D. Complimentary

__________ **28.** A(n) _________ is a plane figure generated around a centerpoint.

__________ **29.** The _________ is 3.1416 times the diameter of a circle.

__________ **30.** A(n) _________ is a line from circumference to circumference not through the centerpoint.

T F **31.** A segment is one-fourth of a circle.

__________ **32.** _________ circles have different diameters and the same centerpoint.

__________ **33.** A(n) _________ is a three-sided polygon with three interior angles.

__________ **34.** A(n) _________ is the portion of a circle set off by a chord.

__________ **35.** A(n) _________ is a many-sided plane figure.

__________ **36.** A(n) _________ is a straight line touching the circumference of a circle at only one point.

__________ **37.** A(n) _________ triangle contains one 90° angle and no equal sides.

A. right
B. isosceles
C. equilateral
D. scalene

__________ **38.** A(n) _________ triangle has three equal angles and three equal sides.

__________ **39.** A(n) _________ is a four-sided polygon with four interior angles.

__________ **40.** The _________ of a triangle is the side upon which the triangle stands.

__________ **41.** A 35′ × 125′ building has an area of _________ sq ft.

__________ **42.** A 50′ × 130′ lot has an area of _________ sq ft.

_____________ **43.** A 9′-6″ × 14′-0″ shed has an area of _________ sq ft.

_____________ **44.** A 4′ × 12′ drywall sheet has an area of _________ sq ft.

_____________ **45.** A triangle with a base of 16″ and an altitude of 25″ has an area of _________ sq in.

Job A

Job A is stocked with 10 sheets of 4′ × 8′, 30 sheets of 4′ × 12′, and 26 sheets of 4′ × 14′ drywall.

_____________ **1.** The total number of sheets stocked is _________.

_____________ **2.** The number of sheets required to do the job if two 4′ × 8′ sheets are left after the job is completed is _________ sheets.

_____________ **3.** The number of sheets that were actually hung if waste accounts for three 4′ × 12′ sheets is _________ sheets.

_____________ **4.** Stocked drywall equals _________ sq ft.

_____________ **5.** Drywall wasted equals _________ sq ft.

Job B

Job B is a cylindrical display building that must be hung with drywall. The building diameter is 80′ and the ceiling height is 12′. There are six 3′ × 7′ doors and no windows. Round answers to the nearest foot.

_____________ **1.** The circumference of the building is _________ ft.

_____________ **2.** The square feet of drywall that must be ordered to hang the walls of the building is _________ sq ft. *Note:* Deduct the door openings and add 10% for waste.

_____________ **3.** The square feet of drywall that must be ordered to hang the ceiling is _________ sq ft. *Note:* Add 10% for waste.

_____________ **4.** The number of 4′ × 12′ sheets that must be ordered for the entire job is _________ sheets.

2 Materials

Heavy- and light-gauge framing materials and metal ceiling framing materials are used to construct interior wall and ceiling drywall systems. Gypsum panels, adhesives, trims, nails, screws, and staples are used along with drywall taping and finishing materials to complete the construction and installation of drywall systems.

GYPSUM PANELS

Gypsum panels (drywall or Sheetrock®) are the most popular wall and ceiling covering material used in residential and commercial construction. *Drywall* is interior surfacing material applied to framing members using dry construction methods such as adhesive or mechanical fasteners. *Sheetrock®* is a brand of gypsum panel developed by the United States Gypsum Company for interior wall and ceiling surfaces. The use of drywall for residential, commercial, and industrial construction has increased for more than half a century. This increase in use is due to the speed and convenience with which drywall can be installed and the minimal cleanup required. Drywall applications have greatly reduced drying time, masking and protection, and amount of cleanup compared to plaster or wetwall applications.

Drywall Manufacture

Drywall sheets consist of a gypsum core surrounded by heavy face and back paper. See Figure 2-1. Drywall sheets are formed by sandwiching semiliquid gypsum slurry between two layers of treated heavy paper. A conveyor carries the gypsum and paper sandwich from the mixer to the drying kilns, where excess moisture is removed. The conveyor carrying the raw drywall sheet products to the drying kilns is long enough to allow the gypsum core to set before being rough cut to length. The raw drywall sheet products are cut into 8′, 9′, 10′, 12′, 14′, and 16′ lengths and packaged for shipment after they have been placed in the drying kilns.

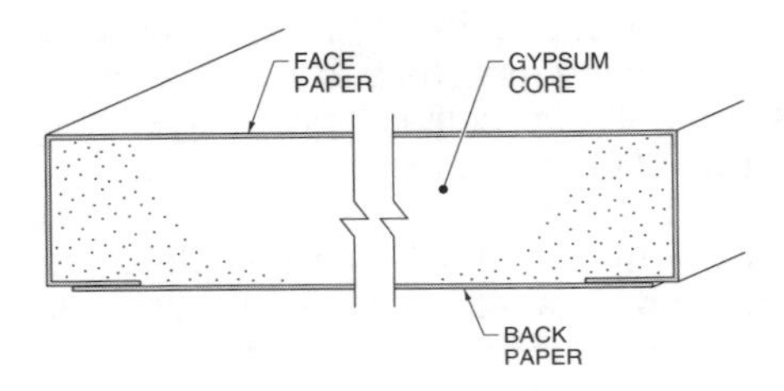

Figure 2-1. Drywall sheets consist of a gypsum core surrounded by heavy face and back paper.

Face and Back Paper

The paper used for the face of regular and fire-resistant drywall sheets is a multilayer paper which provides a smooth surface for finishing and decorating. The back paper is also a multilayer paper, but is unsuitable for finishing. The face and back paper provide strength to the gypsum core.

Gypsum board products were first developed and marketed early in the twentieth century. Early gypsum board products were produced by placing a gypsum mix in approximately 3′ square forms one sheet at a time. Today, gypsum mines and gypsum product manufacturing plants are located throughout the world. These facilities are equipped with machines designed to produce a wide variety of drywall sheet products. Drywall sheet products are available in a wide variety of types, widths, and lengths.

Core Composition

The core composition of drywall sheet products varies according to the use for which the sheets are manufactured. Regular drywall sheets contain a basic core consisting primarily of gypsum, paper pulp, starch, and soap. Fire-resistant drywall sheets contain fiberglass, perlite, or vermiculite, which is added to the basic core materials. Other ingredients are added to the core material when water-resistant drywall sheets are produced. Water-resistant drywall sheets also have a special mildew-resistant, asphalt-impregnated paper used on the front and the back of the sheet.

Drywall

Various drywall sheets are manufactured for use in residential and commercial construction. Drywall sheets include regular, fire-resistant, flexible, moisture-resistant, ceiling and soffit, gypsum sheathing, backer board, coreboard, and decorative drywall. *Gypsum sheathing* is exterior wallboard consisting of a water-repellent gypsum core with a water-repellent paper on face and back surfaces. *Backer board* is drywall installed in a suspended ceiling which serves as an attachment surface for acoustical tile. *Coreboard* is a panel product consisting of a gypsum core encased with strong liner paper, forming a 1″ thick panel. Coreboard is covered with additional layers of drywall panels for use in elevator shafts and laminated gypsum partitions.

Sheet Sizes. Drywall sheets are available in standard thicknesses. See Figure 2-2. The standard width of regular, fire-resistant, moisture-resistant, ceiling and soffit, and decorative drywall sheets is 48″. Widths wider or narrower than 48″ may be obtained from certain manufacturers on a special-order basis. Gypsum sheathing, backer board, and coreboard are manufactured in standard widths of 24″. Regular, fire-resistant, moisture-resistant, ceiling and soffit, and coreboard are available in lengths of 8′, 9′, 10′, 12′, 14′, and 16′. Flexible drywall, gypsum sheathing, backer board, and decorative drywall are available in a standard length of 8′. Most drywall sheets may be obtained in lengths other than those listed on a special-order basis from some manufacturers.

DRYWALL STANDARD THICKNESSES

Type	Thickness*
Standard	1⁄4, 3⁄8, 1⁄2, 5⁄8
Fire-resistant	1⁄2, 5⁄8, 3⁄4
Flexible	3⁄8, 1⁄2
Moisture-resistant	1⁄2, 5⁄8
Ceiling and soffit	1⁄2, 5⁄8
Gypsum sheathing	1⁄2, 5⁄8
Backer board	1⁄2, 5⁄8
Coreboard	1
Decorative	1⁄2, 5⁄8

* in in.

Figure 2-2. Drywall sheets are available in standard thicknesses.

Sheet Edges. Drywall applications normally include paper or fiberglass joint tape and joint (finishing) compound applied to the joints of the drywall sheets. Standard and fire-resistant drywall sheets have tapered side edges to facilitate joint taping and finishing processes. The tapered shape of the side edges provides a finished wall and ceiling surface free from visible joints. A shallow depression is formed when the side edges of two drywall sheets are placed together. This depression allows the tape and joint compound to be applied to the joint while maintaining a flush surface. The top and bottom ends of a drywall sheet are not tapered. See Figure 2-3.

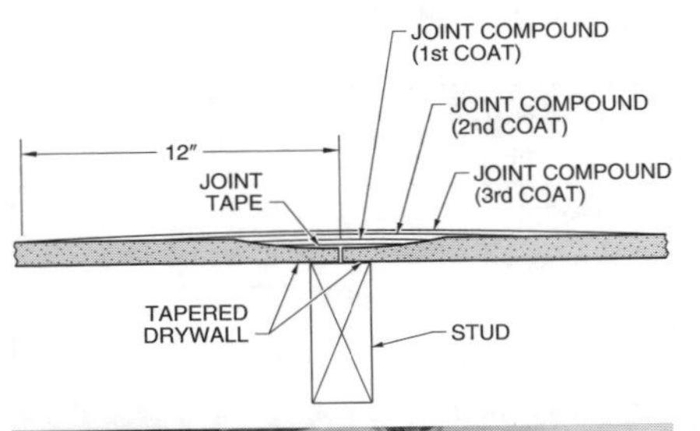

Figure 2-3. Standard and fire-resistant drywall sheets have tapered side edges to allow tape and joint compound to be applied to the joint while maintaining a flush surface.

The joint compound must be built up slightly at the joint when applying tape and joint compound to the end joints. To reduce the visibility of the buildup, the topping coats of joint compound are troweled beyond the joint at least 12″ on each side. When properly applied, the tape and joint compound treatment of the drywall side joints reduce their visibility to an acceptable level.

Demountable Partitions. Drywall sheets are produced for use in a variety of demountable partition systems. A *demountable partition system* is a wall system that uses components designed to be disassembled and reused. Demountable partition systems may be supplied with beveled or squared edges depending on the application. Drywall sheets used for demountable partition systems may also be covered with vinyl or fabric material and are not intended to receive any other finishing.

Most demountable partition systems are designed for the drywall sheets to be installed vertically. This produces a finished wall surface with only the edge joints of the drywall sheet exposed. See Figure 2-4.

Some demountable partition systems use batten strips to conceal the edge joints, others leave the joints exposed. A *batten* is a narrow strip of wood, metal, plastic, or drywall used to conceal an open joint. Some demountable partition systems conceal the vertical joints by inserting them into floor-to-ceiling struts. A *strut* is a member fixed between two other members to maintain a specified distance.

United States Gypsum Company

Figure 2-4. A demountable partition system uses components designed to be disassembled and reused.

Various types, thicknesses, widths, and lengths of drywall sheets are available to meet the requirements of today's construction market. Drywall sheet products are the most versatile and popular material used to cover wall and ceiling surfaces. Specific information related to the various types of drywall products, with their particular application procedures, is found in construction plans and specifications. This information is provided by building designers to ensure that the project is constructed properly.

Drywall Adhesives

An *adhesive* is a substance used to bond two surfaces together. Drywall adhesive provides a method for installing drywall which reduces or eliminates the need for nails, screws, or other fasteners. Drywall adhesives also eliminate the need to fill the depressions created when fasteners are used to attach drywall sheets to framing members. Adhesive-applied drywall sheets reduce the number of nail pops and provide a finished wall surface with fewer blemishes. The application of some prefinished drywall sheets does not permit the use of exposed fasteners. In these applications, adhesives provide the only acceptable method for applying the drywall sheets.

Panel Adhesives. Various panel adhesives are available for standard and prefinished drywall installations. Different panel adhesives are designed for use when installing drywall on wood framing members, metal framing members, and wood or metal framing members. These panel adhesives are packed in two sizes of cardboard tubes. The cardboard tubes are similar to those used for caulking compound. See Figure 2-5.

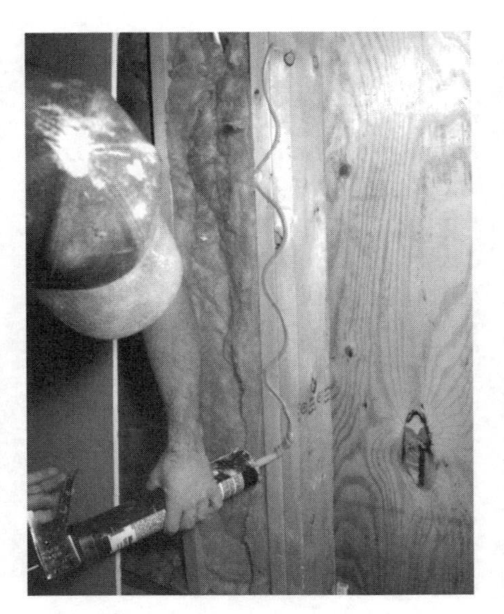

Figure 2-5. Various panel adhesives are available for standard and prefinished drywall installations.

Panel adhesives are used by extending the plunger of a cartridge gun and inserting the tube of adhesive. The plastic tip of the cardboard tube is cut off to provide a ¼″ bead of adhesive. The handle of the cartridge gun is repeatedly squeezed to start and continue the flow of adhesive. Panel adhesive is normally applied to framing members in a zigzag pattern and is only applied to framing members which are covered by one sheet of drywall at a time. The individual drywall sheet must be installed immediately after applying the adhesive. The surface of the adhesive glazes (hardens) and the bond between the framing member and the sheet does not occur if the installation of the sheet is delayed.

When using panel adhesive to apply drywall sheets, temporary bracing or fasteners are required to secure the sheets in place until the adhesive has completely set. This may present a challenge when applying prefinished material where the use of exposed fasteners is prohibited.

Contact Adhesives. Contact adhesives may be used for a variety of drywall applications. Contact adhesive applications require the drywall hanger to position the drywall sheet precisely before it makes contact with the surface being covered. For example, contact adhesives may be used for certain double-layer drywall installations to provide an increase in sound control and fire rating. *Double-layer drywall applications* are applications in which two layers of drywall are installed. Contact adhesive is a popular attachment method when drywall sheets are applied over existing plaster or drywall walls during remodeling work. Contact adhesive requires no temporary bracing or fasteners. A permanent bond occurs when the material being applied makes contact with the substrate.

Contact adhesives may be solvent-based or water-based adhesives. Solvent-based and water-based adhesives are used for a variety of interior applications. They provide excellent bonding of the drywall sheets to the substrate.

Warning: Precautions must be taken when using contact adhesives because solvent-based adhesives release vapors which are mildly toxic and highly flammable. When using solvent-based contact adhesive, workers must wear approved protective breathing masks and the work area must be properly ventilated and free of open flames and sparks. Smoking is not permitted. Any violation of these precautions may lead to serious injury to the workers and costly damage to the structure. Lacquer thinner or other similar solvents are needed to clean tools used to apply solvent-based adhesives.

Water-based adhesives may also be used for a variety of drywall applications. Water-based adhesives use a water-based formula and do not present a potentially dangerous working environment. Another advantage of using water-based adhesives is that tools are easily and quickly cleaned with water.

Joint Compound. Joint compound provides an economical adhesion and may be used to laminate drywall sheets together in a double-layer application. Joint compound is available in dry powder and premixed packaging. See Figure 2-6. The dry powder is mixed with water at the job site. Joint compound does not provide instant bonding and requires a 24-hour drying period when used as an adhesive. Temporary bracing or fasteners must be used to hold the drywall sheets in place until the adhesive has thoroughly dried.

Figure 2-6. Joint compound provides an economical adhesion and may be used to laminate drywall sheets together in double-layer applications.

Most water-based laminating compounds require a similar drying time and the use of temporary bracing or fasteners. Once dried, water-based laminating compounds provide a good bond. Water-based laminating compounds are not recommended for applications where time is the major consideration or where moisture is present.

Lightweight joint compound weighs up to 25% less than standard joint compound and may be used for finishing drywall joints and for patching interior drywall and plaster surfaces.

The adhesive selected normally depends on the application, the allotted completion time, and the information provided on the project plans and specifications. Adhesive-applied drywall sheets, if installed properly, provide a solid, durable wall surface with a minimum number of blemishes.

Drywall Trims

Drywall trims may be trims designed for use on regular drywall applications or special prefinished trims or moldings used with vinyl and fabric covered drywall applications. Regular and special drywall trims are available in a variety of materials, shapes, and finishes. Metal and plastic trims are designed and produced by a variety of manufacturers for use with regular drywall applications. Both provide edge and corner protection and straight line terminations which cover the edges and ends of the drywall sheets. The edges and ends might be exposed without the installation of the trim members. Trim styles available for regular drywall applications include square and bullnose corner bead, L-trim, J-trim, and U-trim.

Drywall trims and accessories are fabricated from light-gauge galvanized steel or PVC plastic. The surface or surfaces of the trim member that lie against the face of the drywall are perforated to allow the joint compound to adhere to the drywall face paper. This forms a strong bond between the trim member and the face of the drywall. The construction plans and specifications indicate the drywall trims to be used, the installation methods to be employed, and the extent to which each trim member is used.

Corner Beads. Drywall corner bead is the most frequently used trim. A *corner bead* is a light-gauge, L-shaped galvanized metal device used to cover and protect the exposed outside corners of drywall. Corner bead is available in square and bullnose styles. Corner bead is coated with several applications of joint compound to provide smooth and straight finished outside corners. The number of coats is indicated in construction specifications. See Figure 2-7.

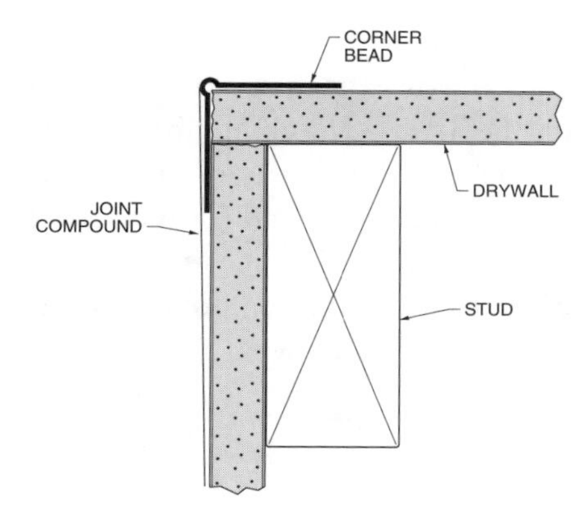

Figure 2-7. A corner bead is used to cover and protect the exposed outside corners of drywall.

L-Trim. L-trim (L-metal) is used to terminate a finished drywall sheet when it butts against another material, such as steel, concrete, masonry, or wood. See Figure 2-8. L-trim may also be used at window and door frames when a flush-finished surface with no casing is desired. L-trim is available in square and bullnose styles.

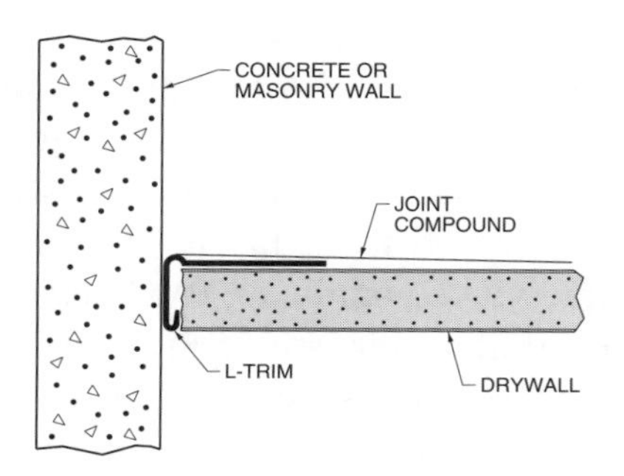

Figure 2-8. L-trim is used to terminate a finished drywall sheet when it butts against another material.

J- and U-Trim. J- and U-trim (J-metal and U-metal) are designed to be used when the edge of the drywall sheet is visible as a finished product. See Figure 2-9. J- and U-trim completely wrap around the edge of the drywall sheet. They are installed with either the perforated surface or the smooth surface visible. Construction plan details indicate J-trim and U-trim installation. J-trim and U-trim are coated with joint compound and provide a flush finish when they are installed with the perforated surface visible. The trims are painted along with the drywall surface when they are installed with the smooth surface visible.

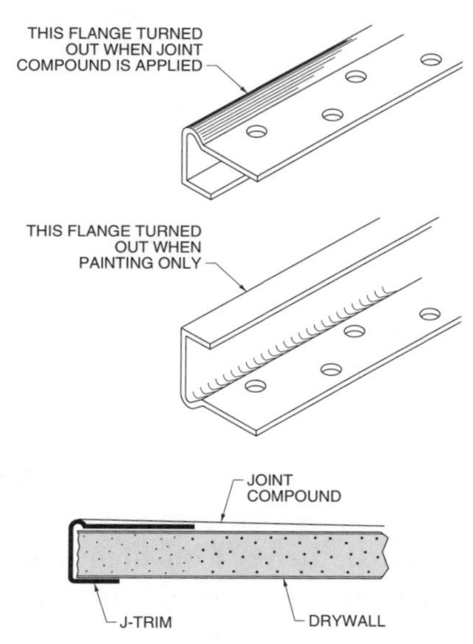

Figure 2-9. J- and U-trim are used when the edge of the drywall sheet is visible as a finished product.

Control-Joint Trim. A *control-joint trim* is a thin strip of perforated metal applied to relieve stress resulting from expansion and contraction in large ceiling and wall surfaces. Control-joint (expansion-joint) trim is designed to be used when the drywall sheets are installed with visible surface control as part of the finished product. See Figure 2-10. Control-joint trim is installed with the perforated surface of the trim applied to the surface of the drywall sheets on both sides of the joint and is coated with joint compound to provide a finished control joint.

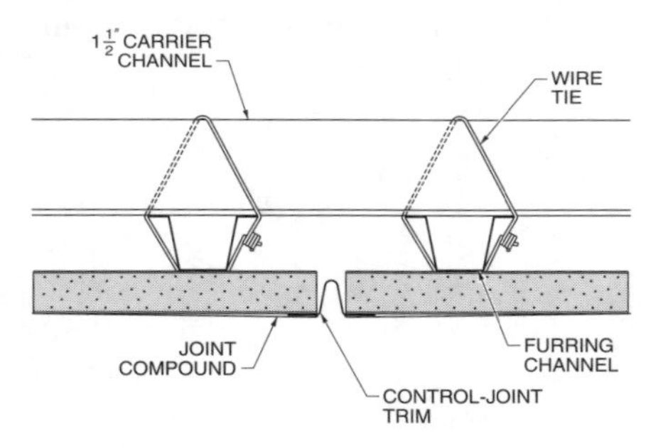

Figure 2-10. Control-joint trim is used when the drywall sheets are installed with visible surface control as part of the finished product.

Reveal Trim. *Reveal trim* is a solid channel-shaped strip of metal with perforated surfaces on both sides. Reveal trim (Fry® trim) is designed to be used when drywall sheets are installed with visible surface reveals as part of the finished product. A *reveal* is a drywall joint that is open to view. Reveal trim is installed with the perforated surface of the trim applied to the surface of the drywall sheets on both sides of the space provided for the reveal. See Figure 2-11. Reveal trim is coated with joint compound to provide a finished reveal.

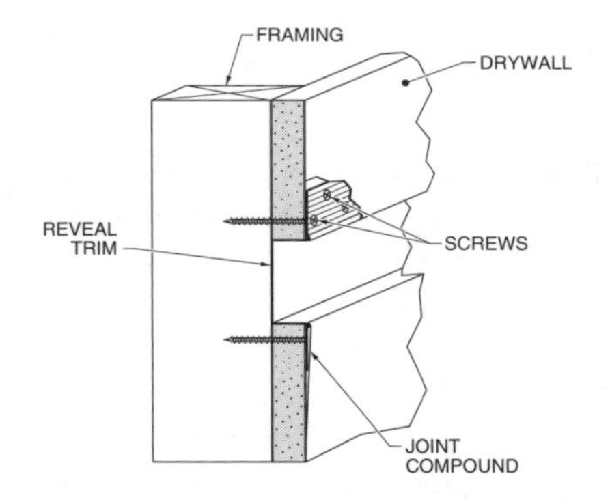

Figure 2-11. Reveal trim is designed to be used when drywall sheets are installed with visible surface reveals as part of the finished product.

Prefinished Drywall Trim. Prefinished drywall trims (moldings) serve a similar purpose as regular drywall trims, except they are designed for use with vinyl or fabric covered drywall or other prefinished panels. Prefinished drywall or panel trim is fabricated from rigid vinyl, aluminum, PVC plastic, or light-gauge steel. Prefinished drywall trim is designed to complete the finished installation and may be colored to complement the material used to cover the prefinished panels.

A wide variety of trim and accessories are available for use with prefinished panel applications. They include corner trim, F-corner trim, J-trim, and batten trim. Corner trim is used for concealing drywall sheet edges when they form an outside corner. F-corner trim is an alternate trim member used to form an outside corner. J-trim is used where drywall sheets abut a foreign material, such as masonry, concrete, wood, etc. Batten trim is used to conceal the joints between two drywall sheets. See Figure 2-12. These trims are normally installed in a manner that provides for all fasteners (nails or screws) and exposed edges of the prefinished drywall sheets to be covered.

Drywall trims and moldings, when properly installed, enhance the appearance and durability of a finished wall surface. The drywall hanger must ensure that all drywall trims are installed properly because the trim members determine the final shape of the finished product. All trim members must be installed straight and true.

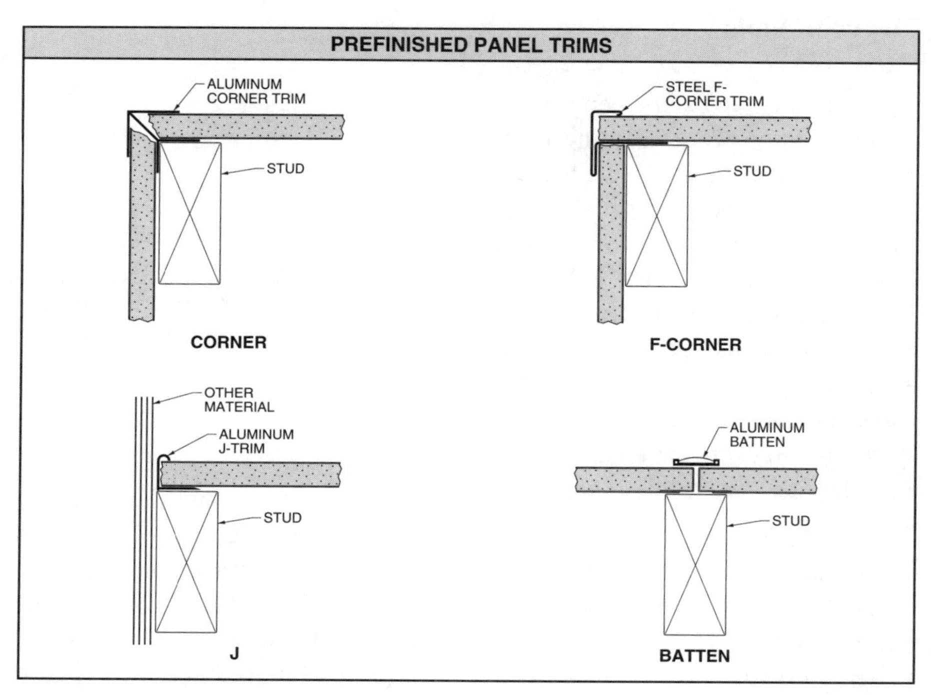

Figure 2-12. Prefinished panel trims include corner, F-corner, J-, and batten trim.

Carelessly applied, bent, or defective drywall trims produce finish lines that are crooked and must be replaced prior to taping. Such installations produce results that may cause the entire project to be unacceptable to the owner. Moldings must be free from blemishes or other damage and installed with true-cut and tight-fitting joints. Just as drywall trims can determine the appearance of the finished product, moldings that are properly or improperly installed add to or detract from the overall appearance of the finished product.

Drywall Nails

Special nails are used when drywall is installed on wood framing members. Because most drywall installations are taped and finished, the nails used must serve two purposes. First, they must have adequate holding power to ensure that the drywall remains securely fastened-in-place. Second, the nail must have a head designed to securely hold the drywall without cutting into the face paper. See Figure 2-13.

The drywall nail selected must be the correct length to provide 1″ of penetration into the wood framing member. Drywall nails that are driven into wood framing members to the proper depth help reduce nail pops. Nail pops are caused in part by the space that is created between the back of the drywall sheet and the face of the wood framing member as the wood dries and shrinks.

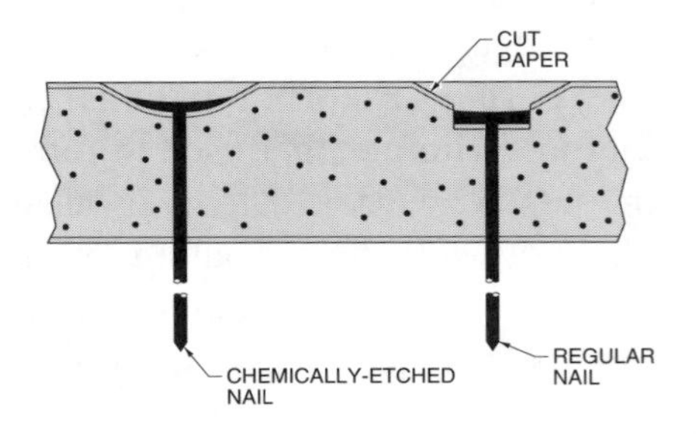

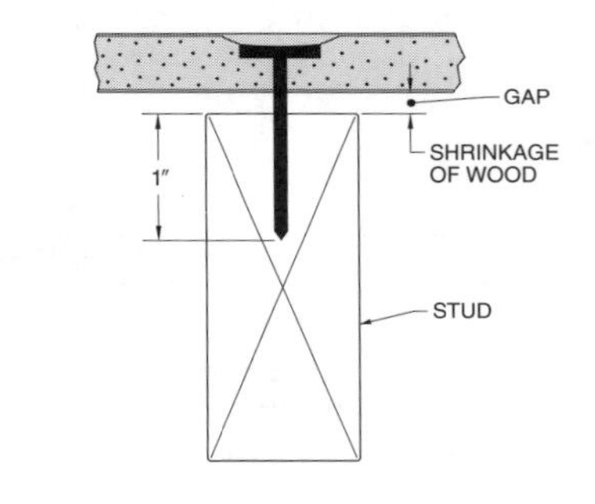

Figure 2-13. Drywall nails must have adequate holding power and a head designed to securely hold the drywall without cutting into the face paper.

Nail Pops. *Nail pops* are blemishes caused when drywall nail heads force the finishing material past the surrounding surface and become exposed. See Figure 2-14. Nail pops can be corrected by driving the popped nail back into place with a nail set and filling the hole with putty or drywall topping compound. A *nail set* is a small steel punch-like tool used to set finish nails below the surface of a trim member. The nail set must be centered on the head of the nail and struck with a hammer to drive the nail into the framing member. If done properly, the nail head is driven below the surface, leaving a small hole which is easily filled. Nail pops may also be corrected by driving another nail beside the one that has popped and refinishing the surrounding drywall surface.

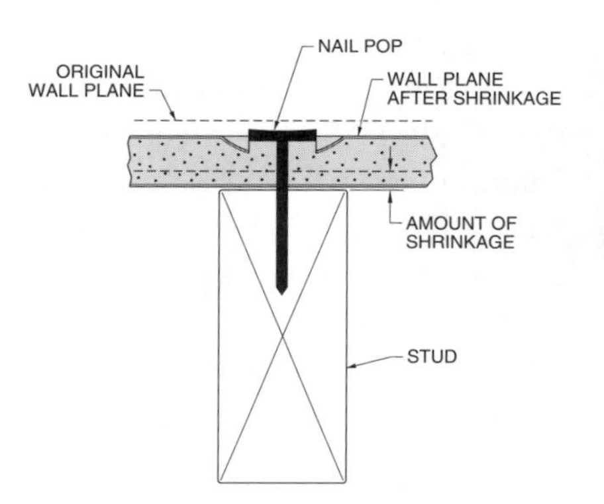

Figure 2-14. Nail pops are blemishes caused when drywall nail heads force the finishing material past the surrounding surface and become exposed.

Shrinkage of wood framing members is a common occurrence and must be compensated for when installing drywall. Nails that penetrate wood framing members more than 1″ increase the incidence of nail pops in the finished drywall. Nails driven into wood framing members push the drywall sheets off the face of the wood as the wood dries and shrinks. This creates a gap between the back of the drywall sheet and the face of the wood framing member. This gap allows the nail head to pop through the topping compound when the surrounding drywall is pushed against the framing member.

Firestop® X gypsum panels have a core that contains special additives to enhance the integrity of the core under fire exposure.

Chemically-Etched Nails. Chemically-etched nails are specified on 95% of all single-layer drywall installations applied on wood framing members when the drywall is to be taped and finished. Chemically-etched drywall nails have superior holding power and the cupped head greatly reduces cutting the face paper of the drywall when the nail is driven the proper depth with the proper dimple. See Figure 2-15.

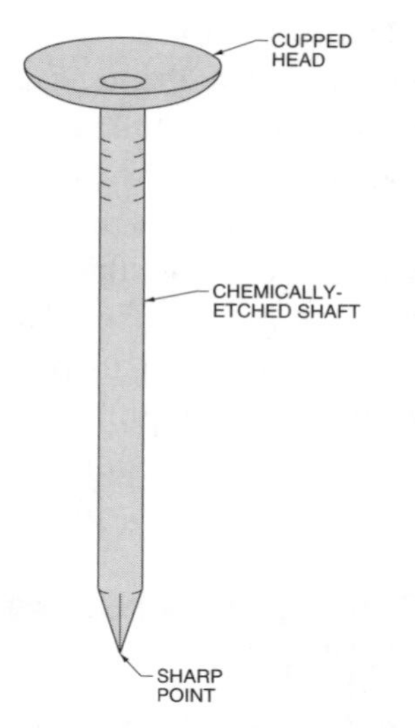

Figure 2-15. Chemically-etched drywall nails have superior holding power and the cupped head greatly reduces cutting the face paper of drywall sheets.

Chemically-etched nails allow a smooth dimpled effect where regular box nails would cut the face paper of drywall sheets. Chemically-etched nails are available in a variety of lengths from 1⅜″ to 3½″. The introduction of chemically-etched nails has significantly reduced the incidence of nail pops.

Cement-Coated Cooler Nails. Cement-coated cooler nails may be used for fire-rated assemblies and special drywall applications such as installing multiple layers. Cement-coated cooler nails may also be used for nailing drywall that is to be covered with wood paneling, acoustical tile, or other materials. They are available in a variety of lengths from 1⅜″ to 3½″.

Color Pins. Color pins (colored nails) may be used when applying vinyl or fabric drywall sheets onto wood framing members. Color pins are designed for this type of application and are coated with a special paint to match the material used to cover the drywall. The installation of prefinished drywall sheets normally requires the use of panel adhesive to bond the back of drywall sheets to the framing members. Color pins are used to hold the sheets in firm contact with the framing members until the adhesive has fully dried. When properly installed, color pins blend with the material covering the drywall sheets and provide an attractive finished appearance.

Always select and use the correct length of drywall nail for the application. The nail must penetrate the wood framing member as near to 1″ as possible. This depth ensures that the gap created between the back of the drywall sheet and the framing member is kept to a minimum when the framing member dries and shrinks. Longer nails with greater penetration into the wood framing member create a larger gap that is unacceptable.

Drywall Screws

Drywall screws of various lengths are used to attach drywall sheets to wood, heavy-gauge metal framing, and light-gauge metal framing members. See Figure 2-16. Drywall screws are designed with self-drilling and self-tapping points. They have coarse threads which provide increased holding power and bugle-shaped heads that compress the drywall face paper without cutting it as the screw is driven. The desired results are achieved using fewer fasteners because drywall screws have superior holding power compared to nails.

DRYWALL SCREWS		
Fastener	**Sizes***	**Fastening Applications**
Bugle Head	1, 1⅛, 1¼, 1⅝	Single- and double-layer gypsum bases to steel studs, metal furring, and resilient channel
S Bugle Head	1¼, 1⅞, 2¼, 2⅝	Intermediate layers and face layer gypsum base to steel runners in caged beam fireproofing
S-12 Bugle Head	1, 1¼, 1⅝, 1⅞, 2⅜, 2⅝, 3	Gypsum sheathing and gypsum base to No. 20 gauge or thicker steel studs in curtain wall assemblies
S Pan Head	7⁄16, ⅜	Steel studs to steel runners
S-12 Pan Head	⅜, ½	Steel studs to runners, metal door frame
S-12 Low-Profile Head	½	Steel-to-steel attachment up to No. 12 gauge total in curtain walls and steel framing
W Bugle Head	1¼	Single-layer gypsum base to wood framing. Resilient channels to wood framing
G Bugle Head	1½	Face-layer gypsum to base-layer gypsum in laminated partitions

* in in.

Figure 2-16. Drywall screws are used to attach drywall to wood, heavy-gauge metal framing members, light-gauge metal framing members, and for securing other framing members.

Drywall screws are designed to be driven with an electric or pneumatic screwgun. They are available in lengths from 1″ to 3″. This variety of lengths accommodates a wide range of drywall thicknesses and number of layers. Drywall screws are chemically treated to resist rusting when the drywall finishing compound is applied. Care must be exercised to keep them stored in a dry location. Smaller self-drilling, self-tapping screws are also available for attaching various metal framing members in a metal framing assembly. These screws are also driven with an electric or pneumatic screwgun.

Staples

Staples are normally used only for attaching the first layer of drywall to wood framing members in a double-layer installation. The outer layer is nailed or screwed to the wood framing members or laminated to the first layer of drywall. Staples are also used to attach drywall to wood ceiling framing when the finished ceiling receives stick-on acoustical tile. Staples are unsuitable for single-layers or the outer layer of double-layer applications because of inconsistency in penetration. Fasteners for the final layer of drywall are recessed to provide a smooth, finished surface.

Wide-crown staples are used when installing drywall. The force exerted on the staple is adjusted by varying the air pressure supplied to the staple gun. The crown is driven tight against the face of the drywall with a minimal amount of cutting into the face paper when the air pressure is properly adjusted. See Figure 2-17. Staples provide a fast and economical way of fastening drywall sheets to framing members.

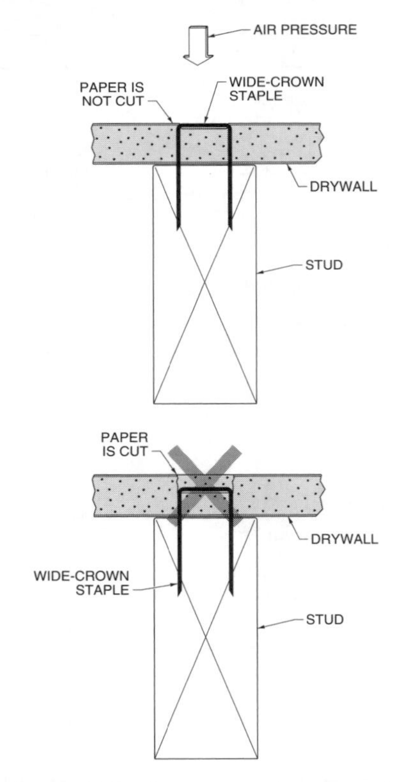

Figure 2-17. The crown is driven tight against the face of the drywall with a minimal amount of cutting into the face paper when the air pressure is properly adjusted.

HEAVY-GAUGE METAL FRAMING MATERIALS

Noncombustible heavy-gauge metal framing components have been developed for interior and exterior wall and ceiling assemblies. These members may be used in place of wood studs and ceiling joists. Heavy-gauge metal and wood stud walls are assembled in a similar manner. Wood-framed walls use studs and plates. Heavy-gauge metal framing systems use studs and tracks or runners. Heavy-gauge metal studs and tracks are channel-shaped and are formed from shop primed and galvanized sheet steel. See Figure 2-18.

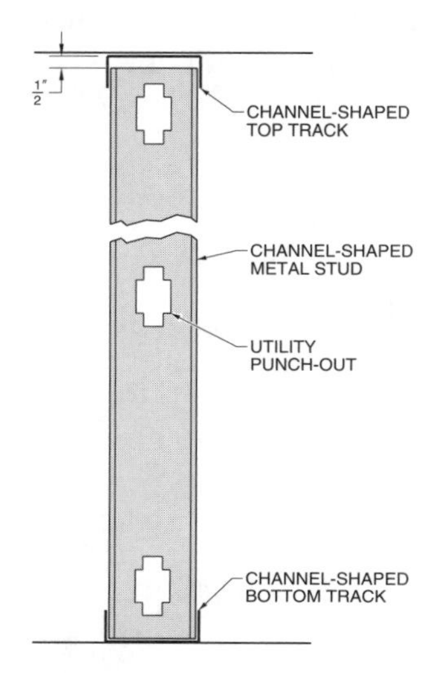

Figure 2-18. Heavy-gauge metal studs and tracks are channel-shaped and are formed from shop primed and galvanized sheet steel.

Utility Holes

Manufacturers of heavy-gauge metal studs provide holes punched at 24″ intervals the entire length of the stud. Some manufacturers may provide punch-outs spaced more frequently than 24″. Stud punch-outs are used for utilities, such as plumbing and electrical services, and to facilitate the installation of internal wall bracing. Heavy-gauge metal stud track is also supplied in a channel-shape design to receive the metal studs which fit into the open face of the channel.

Heavy-Gauge Metal Studs

Heavy-gauge metal studs are available in standard steel thicknesses of No. 18, No. 16, No. 14, and No. 12 gauge. Standard studs are available in widths of 2½″, 3¼″, 3⅝″, 4″, 6″, 8″, 10″, and 12″, with flange widths of 1″, 1¼″, 1⅜″, and 1⅝″. Standard stud lengths range from 8′ to 24′, but studs are available in custom lengths that may be ordered from most manufacturers. Additional widths may be available from certain manufacturers on a special-order basis. Heavy-gauge metal stud tracks are available in all stud widths to accommodate the stud being used. Metal stud tracks may be purchased in standard and deep-leg styles in lengths of 10′, 12′, and 20′, or in nonstandard lengths on a special-order basis. See Figure 2-19.

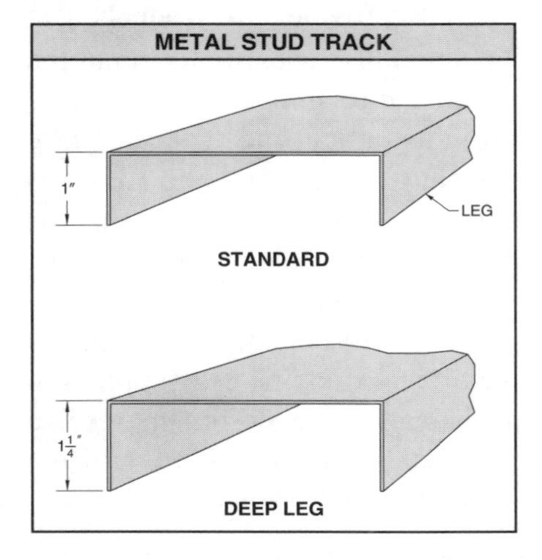

STUD TRACK STANDARD SIZES				
Stud Track	Length*	Width**	Gauge	Leg Length**
Standard	10	2½, 3⅝, 4, 6	25, 22, 20, 18, 16, 14	1
Deep Leg	10	2½, 3⅝, 4, 6	25, 22, 20, 18, 16, 14	1¼

* in ft
** in in.

Figure 2-19. Metal stud tracks are available in standard and deep-leg styles of various lengths, widths, and gauges.

Heavy-gauge metal framing members are cut to length using an electrical- or gasoline-powered cutoff saw. *Note:* The cut edges of metal studs and tracks may be sharp and cause severe injury to the person handling the framing components.

Bracing

Walls framed with heavy-gauge metal studs may require additional bracing. The bracing is installed using 1½″, No. 16 gauge cold-rolled steel channels, or sections of the same size and gauge studs or track used to frame the wall. The construction plans and specifications contain information regarding the type, location, and method of attachment. Bracing may be required when the wall studs are subjected to above-average wind or seismic loading and when the height of the walls exceeds the design limits for unbraced and unbridged framing. The addition of wall bracing increases the rigidity of the wall and improves the stability of the entire assembly. See Figure 2-20.

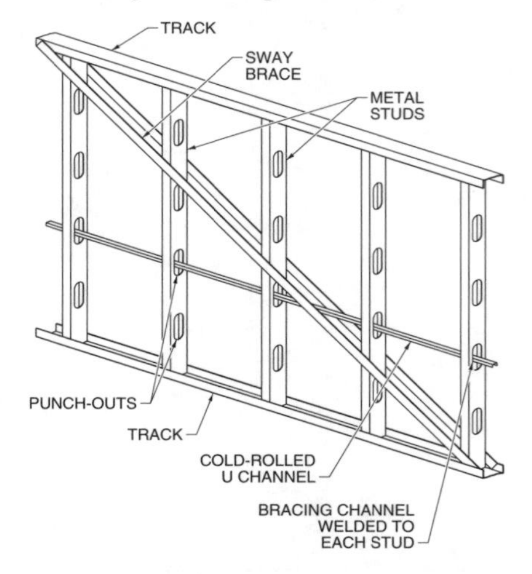

Figure 2-20. Wall bracing increases the rigidity of the wall and improves the stability of the entire assembly.

LIGHT-GAUGE METAL FRAMING MATERIALS

Light-gauge metal framing members have been developed for interior partitions which require noncombustible framing assemblies and may be used in place of wood framing members. Although metal and wood framing members are assembled in a similar manner, light-gauge metal framing has several unique characteristics. Wood framing partition systems use studs and plates. Light-gauge metal framing partition systems use studs and tracks (runners). Metal studs and tracks are channel-shaped and made from galvanized steel. See Figure 2-21.

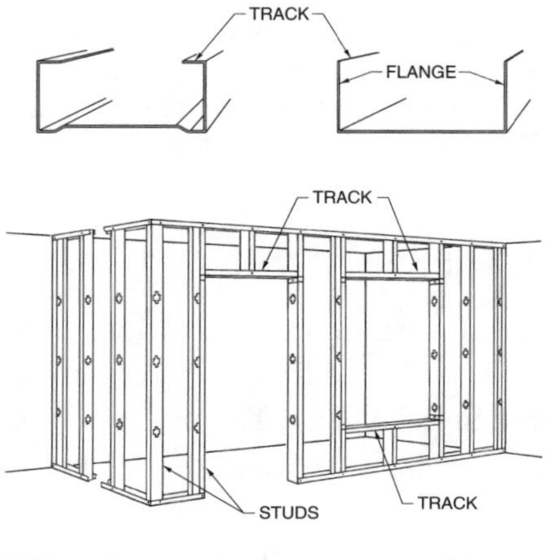

Figure 2-21. Light-gauge metal framing partition systems use studs and tracks (runners).

Utility Holes

Manufacturers of light-gauge metal studs provide punch-outs every 24″ over the full length of the studs. The punch-outs are used for utilities, such as plumbing and electrical services, and to accommodate the installation of internal partition bracing. Service installers may need to use a hole punch if the punch-outs provided do not match the location needed. The channel-shaped track is shaped similar to the studs, and is designed to receive the metal studs which fit into the open face of the channel. See Figure 2-22.

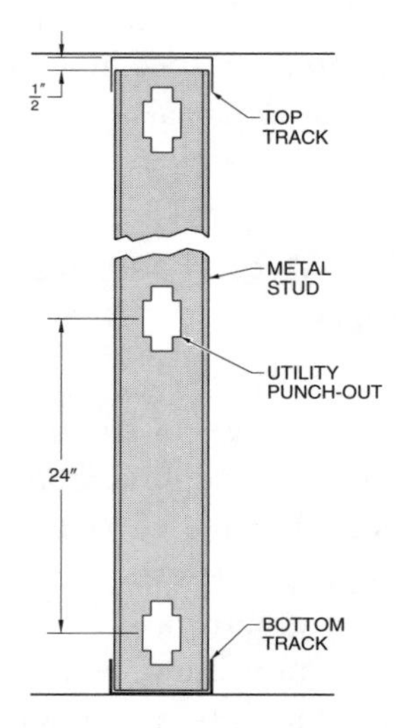

Figure 2-22. Manufacturers of light-gauge metal studs provide holes punched every 24" the full length of the studs.

Light-Gauge Metal Studs

Light-gauge metal studs are available in thicknesses of No. 25 and No. 20 gauge. Standard widths for No. 25 gauge components are 1⅝″, 2½″, 3½″, 3⅝″, 4″, 6″, and 8″. Standard widths for No. 20 gauge components are 2½″, 3½″, 3⅝″, 4″, 6″, 8″, and 10″. Both widths are available in standard lengths of 8′ to 24′. Additional widths and lengths may be available from some manufacturers on a special-order basis. Light-gauge metal stud track is available in the same widths as the stud size being used. It is provided in standard and deep-leg styles, in standard lengths of 10′, 12′, and 20′. Stud track may also be available in special lengths on a special-order basis.

Cutting Metal Framing Members

Light-gauge metal framing members are manufactured from No. 25 and No. 20 gauge galvanized steel. No. 20 gauge members must be cut using an electrical- or gasoline-powered cutoff saw. No. 25 gauge members are easily cut with aviation snips. The stud and track is first cut by cutting the two flanges (sides of the channel shape). See Figure 2-23. The stud or track is bent 90° to open the cut channel flanges and a cut is made across the back of the studs or tracks.

Warning: The cut edges of metal studs and tracks are very sharp and can cause severe injury to the drywall hanger or framer if care is not taken during handling.

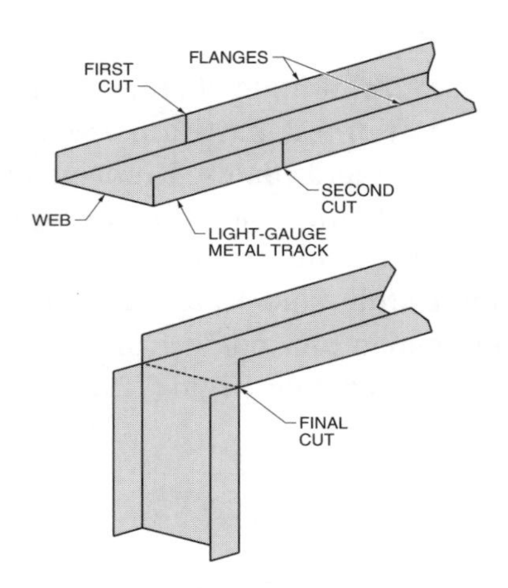

Figure 2-23. No. 25 gauge stud and track members are cut by cutting the flanges and bending the stud or track 90° to open the cut channel flanges.

Bracing

Metal stud framed partitions may require the addition of continuous bracing channels. The bracing is installed using ¾″ and 1½″, No. 16 gauge cold-rolled steel channels. The plans and specifications provide information regarding the type, location, and method of attachment.

Bracing may be required when the partition studs are required to carry above-average loads or the height of the partition exceeds the design limits for unbridged studs. No. 16 gauge cold-rolled channels may be cut with short-nose aviation snips and are placed in the special notches provided by the stud punch-outs. Once the bracing channels are in place, they are securely tied with wire or welded to each stud. See Figure 2-24. Bracing provides additional rigidity and increased stability to the partition.

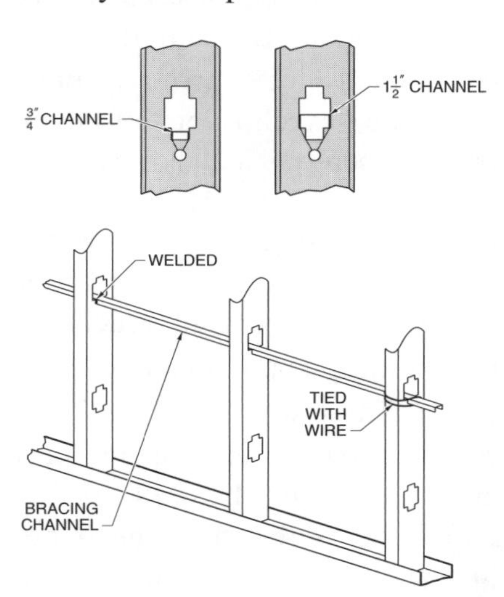

Figure 2-24. Bracing channels are tied with wire or welded to each stud.

Shaftwall Framing

Special metal framing parts are also used for constructing shaftwalls and similar rated enclosures requiring one-hour, two-hour, or longer fire-resistive ratings. A *shaftwall* is a rated enclosure which encloses elevators, air ducts, plumbing pipes, electrical wires, or other items which pass through the floors of a high-rise building. The framing members, drywall, and accessories used are manufactured as complete systems.

The complete shaftwall system includes track, studs, coreboard, and drywall sheets used for the face layer. See Figure 2-25. Shaftwalls are rated assemblies which carry an Underwriters Laboratories Inc.® (UL®) fire test number. The entire assembly is constructed with components supplied by the manufacturer applying for the approval when the test is conducted to determine the UL® rating. For this reason, shaftwall components from different manufacturers shall not be mixed when constructing rated enclosures.

Adhesives for drywall applications meet ASTM C557 standards and are available in 29 oz cartridges for hand or powered gun applications.

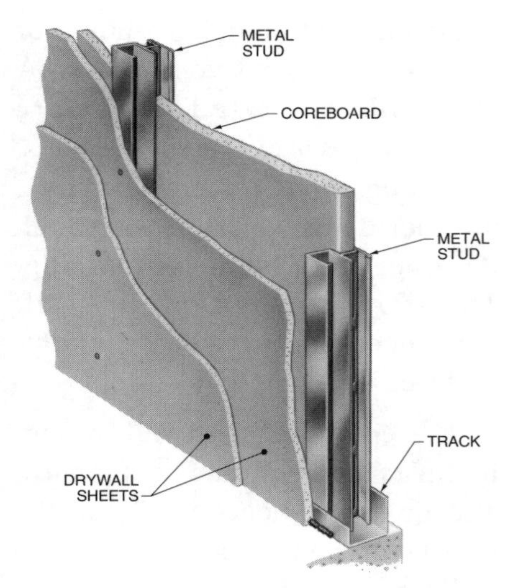

United States Gypsum Company

Figure 2-25. The complete shaftwall system includes track, studs, coreboard, and drywall sheets used for the face layer.

METAL CEILING FRAMING MATERIALS

Metal ceiling framing material provides a skeletal structure to which finished ceiling material is fastened. Several systems are available that may be used when the interior or exterior ceiling is suspended from the overhead structure of the building. Suspended ceilings can create special effects while concealing the ductwork, piping, and other mechanical equipment, as well as the structural components of the building. Suspended ceiling framing must be lightweight yet sturdy enough to support the finished ceiling material.

Suspended Ceiling Framing Materials

The most common suspended ceiling system, other than suspended grid and acoustical tile, is a metal framing system using drywall sheets as a covering. This suspended ceiling system is constructed from furring channels secured to 1½″ cold-rolled steel carrier channels with wire ties, clips, or self-tapping framing screws, and covered with one or more layers of drywall. See Figure 2-26. In this framing system, No. 8 gauge hanger wires are attached to the overhead structure of the building approximately 48″ on center. These wires hold 1½″ cold-rolled carrier channels in place, and the furring channels are tied, clipped, or screwed to the bottom of the 1½″ carrier channels at 16″ or 24″ on center.

Hanger Wire. *Hanger wire* is the wire that supports a suspended member. Hanger wire is supplied in rolls or bundles. Hanger wire bundles contain precut wires in lengths from 4′ to 20′.

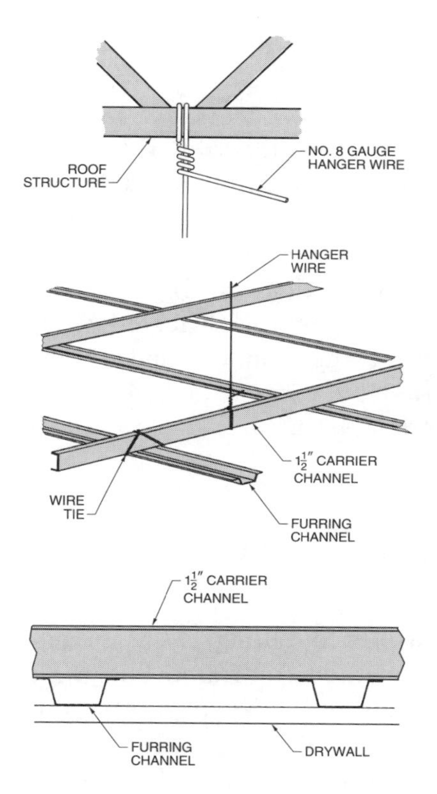

Figure 2-26. A metal framing system using drywall as a covering is the most common suspended ceiling system other than suspended grid and acoustical tile.

Carrier Channels. *Carrier channels* are the main supporting members of a suspended ceiling system to which furring channels are attached. Cold-rolled steel carrier channels may be cut with a power saw, hacksaw, a pair of short-nose aviation snips, or nippers. The procedure for cutting carrier channels by hand is similar to the procedure for cutting light-gauge metal framing members, except that after the flanges of the channel are cut, the channel is bent back and forth until it breaks apart at the point where the flanges of the channel were cut. Cold-rolled steel carrier channels are available in 20′ lengths.

Joint compounds are available in ready-mixed or powder forms. Ready-mixed compounds are opened and used. Powder compounds require the correct amounts of powder and water for best results.

Furring Channels. *Furring channels* are metal channels fastened to a structural surface to provide a base for fastening finish material. Furring channels are supplied in lengths of 12′, 16′, or 20′. Furring channels may also be cut with a power saw, hacksaw, or aviation snips. When cutting furring channels by hand, the two flanges are cut, the channel is bent approximately 90°, and the back is cut. See Figure 2-27.

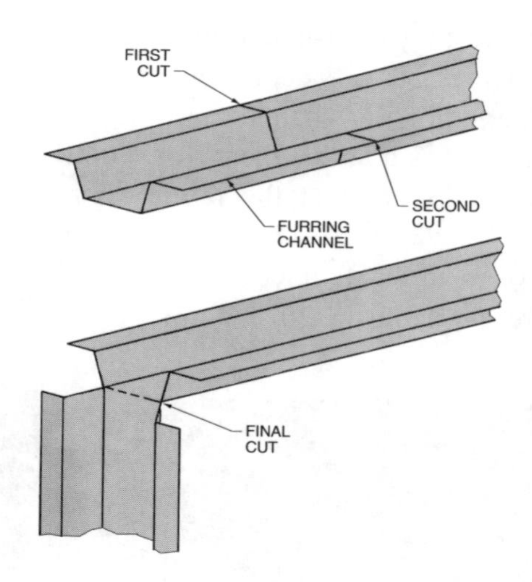

Figure 2-27. When cutting furring channels by hand, the two flanges are cut, the channel is bent approximately 90°, and the back is cut.

A suspended ceiling consisting of metal framing covered with drywall sheets provides a relatively inexpensive method for concealing mechanical equipment, piping, conduit, and the building structure, as well as a finished surface which can be shaped and decorated in a variety of ways. Many different shapes and effects have been created using this type of suspended drywall ceiling. For example, slopes and curves can be created by varying the length of the hanger wires supporting the carrier channels. The sides and ends of suspended drywall ceilings may be curved or angled or have a combination of curves and angles.

Acoustical Ceiling Drywall Grid

Several acoustical tile grid manufacturers produce suspended ceiling grid systems similar to those used for supporting acoustical tile. Acoustical ceiling drywall grid framing is fabricated from interlocking members and is supported by wires attached to the overhead structure.

Joist-Framed Ceiling

Joist-framed ceiling is an alternative method for framing drywall ceilings which incorporates the same principles used when framing metal stud partitions. In a joist-framed ceiling, the metal stud tracks are attached to the walls or partitions on both sides of the room. The metal studs are placed in the track and attached at the top and bottom to complete the ceiling framing assembly. Drywall sheets are attached to the bottom or top and bottom of the framing, depending on the rating, to complete the ceiling assembly. The punch-outs in the metal studs may be threaded with $1\frac{1}{2}''$ carrier channels which are connected to the overhead structure with No. 8 gauge wires when the unsupported length of the studs forming the ceiling exceeds the design limits.

Joist-framed ceilings are often used to create fire-rated public corridors (tunnel corridors) which provide utility space above the ceiling while maintaining the required one- or two-hour fire rating for the interior of the corridor. Depending on the rating requirements of tunnel corridors, one or two layers of drywall may be installed on the top and bottom of the framing members.

Metal Framing Safety Precautions

Precautions must be followed regardless of the type, manufacturer, or style of metal framing system used. These precautions include:

- Exercise care when handling any metal framing material. Cut ends and edges are sharp and can cause serious injury to anyone carelessly handling metal framing components.
- Exercise care to avoid serious cuts to various body parts when cutting No. 25 gauge metal framing components with snips. The cutting action places the hands and fingers in close proximity to the sharp metal being cut and serious injury can result.
- Exercise caution when using power cutoff saws. Electric cords and connections must be in good working condition and properly connected. Engine oil level must be continually checked if using a gasoline-powered model. Gasoline must be properly stored in an approved container and care must be exercised when placing gasoline in the fuel tank. Do not allow gasoline to drip on a hot engine manifold.
- Never operate a cutoff saw without the blade guard in place and operating properly. The blade guard must never be wedged open or removed.
- Eye and ear protection must be worn and gloves are recommended while operating a power cutoff saw.
- Avoid stockpiling metal track, studs, and accessories in locations of heavy foot, cart, or vehicle traffic. Metal framing components are easily bent or damaged by drywall carts, forklifts, scaffolds, or other equipment. Metal framing track, studs, and accessories which are accidentally damaged must be discarded or cut up for smaller pieces. This adds unnecessary cost to the framing operation.
- Metal framing and trim components stockpiled on the job site

must be stacked off the ground away from moisture and, if stored outside the building, covered to protect them from inclement weather. Excess moisture causes rust on the surface of the metal framing members. Excessive rust on metal framing and trim members may result in the components being rejected by the building inspector.

- Keep all work areas and material stockpile areas clean. All scrap metal pieces must be removed to a suitable disposal container as they accumulate. Serious injury can result from slipping or tripping over scrap material.

DRYWALL TAPING AND FINISHING MATERIALS

Drywall finishing materials include paper and fiberglass joint tape, joint taping compounds, and joint topping compounds. Manufacturers obtain sound and fire test results by using a combination of their products as a complete system. As a result, products from different manufacturers may not be mixed when constructing rated wall and ceiling assemblies. For best results, refer to building plans and specifications for the specific job requirements.

Paper Joint Tape

Paper joint tape is fabricated from paper and fiberglass mesh. Paper joint tape is produced from a high-strength reinforced paper which is specially treated for embedding in the joint compound. See Figure 2-28. Paper joint tape is supplied in rolls 500′ long and $2\frac{1}{16}''$ wide. It is normally used for production taping in both commercial and residential construction. Paper joint tape is normally installed with a drywall taping machine (Bazooka®).

AMES Taping Tool Systems Inc.

Figure 2-28. Paper joint tape is fabricated from paper and fiberglass mesh.

Fiberglass Joint Tape

Fiberglass joint tape is produced from fiberglass mesh and is supplied with adhesive applied on one side. See Figure 2-29. It is ideal for patching and small drywall projects when the use of a Bazooka® is impractical. Fiberglass joint tape can be cut with a utility knife or a 6″ taping knife with a sharp outer edge. Fiberglass joint tape is installed by pressing the adhesive side onto the surface of the drywall joint. Once installed, the fiberglass joint tape is coated with joint compound.

Hyde Tools

Figure 2-29. Fiberglass joint tape is produced from fiberglass mesh and is supplied with adhesive applied on one side only.

All-Purpose Joint Compound

All-purpose joint compound (taping mud) is available from most manufacturers of gypsum wallboard products. Some independent manufacturers also produce a line of drywall finishing materials. All-purpose joint compound may be used to embed the drywall tape or as a finishing coat. All-purpose joint compound has ingredients which reduce shrinkage as it dries and make it easier to sand after it has dried. Some drywall tapers prefer to use only all-purpose drywall joint compound, while others only use it for embedding the tape.

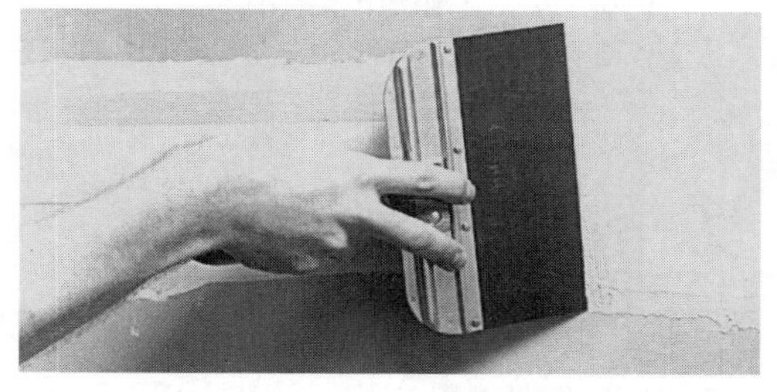

National Gypsum Company

Joint topping compound may be applied with a 12″ taping knife.

Joint Compound

Joint compound is applied directly to the surface of the drywall prior to the placement of joint tape. Joint compound contains an adhesive ingredient which makes it ideal for embedding joint tape. Joint compound may appear slightly more buttery than all-purpose joint compound, and works well in a Bazooka®.

Joint Topping Compound

Joint topping compound is designed to be applied as the second and third coats in the taping and finishing process. Joint topping compound does not contain the adhesive ingredients which are a part of the joint compounds and must not be used for embedding joint tape. Joint topping compound may appear more buttery than all-purpose joint compound and has better sanding characteristics. It is applied with a skim box or by hand using various width taping knives.

Quick-Setting Joint and Topping Compound

Quick-setting joint and topping compound, supplied in powder form, is designed to be applied when there is insufficient time for the standard taping and finishing process. Quick-setting joint and topping compound contains ingredients which cause it to harden within 30 min, 60 min, or 90 min of being mixed with water. It may be used for embedding drywall tape and the successive finishing coats. It appears less buttery than all-purpose and joint compounds and has poor sanding characteristics.

Other drywall finishing products are also available. Refer to construction plans and specifications for the type and manufacturer of the taping and finishing materials that are required for each project.

Drywall is the most popular wall and ceiling covering material used in residential and commercial construction.

Materials

Trade Competency Test

Name Date

Materials

_____________ **1.** A(n) _________ system is a wall system that uses components designed to be disassembled and reused.

_____________ **2.** A(n) _________ is a narrow strip of wood, metal, plastic, or drywall used to conceal an open joint.

_____________ **3.** _________ is drywall installed in a suspended ceiling which serves as an attachment surface for acoustical tile.

A. Gypsum sheathing
B. Coreboard
C. Backer board
D. neither A, B, nor C

_____________ **4.** A(n) _________ is a light-gauge, L-shaped galvanized metal device used to cover and protect the exposed outside corners of drywall.

_____________ **5.** _________ is exterior wallboard consisting of a water-repellent gypsum core with a water-repellent paper on face and back surfaces.

_____________ **6.** _________ is used to terminate a finished drywall sheet when it butts against another material, such as steel, concrete, masonry, or wood.

_____________ **7.** A(n) _________ is a thin strip of perforated metal applied to relieve stress resulting from expansion and contraction in large ceiling and wall surfaces.

T F **8.** Coreboard is a panel product consisting of a gypsum core encased with strong liner paper, forming a 1″ thick panel.

_____________ **9.** _________ are blemishes caused when drywall nail heads force the finishing material past the surrounding surface and become exposed.

T F **10.** Drywall nails are selected to provide 1½″ of penetration into the wood framing member.

__________ **11.** A(n) _________ is a rated enclosure for elevators, stairways, or utilities, such as electrical, plumbing, and air conditioning ductwork.

__________ **12.** _________ wire is the wire that supports a suspended member.

__________ **13.** _________ channels are the main supporting members of a suspended ceiling system to which furring channels are attached.

__________ **14.** Regular drywall sheets contain a basic core consisting primarily of _________.

A. gypsum
B. starch
C. paper pulp
D. A, B, and C

T F **15.** Carrier channels are metal channels fastened to a structural surface to provide a base for fastening finish material.

__________ **16.** The standard width of regular, fire-resistant, moisture-resistant, ceiling and soffit, and decorative drywall sheets is _________″.

T F **17.** Drywall sheets used for demountable partitions systems may be covered with vinyl or fabric material and are not intended to receive any other finishing.

__________ **18.** A(n) _________ is a substance used to bond two surfaces together.

__________ **19.** _________ drywall applications are applications in which two layers of drywall are installed.

__________ **20.** _________ trim is designed to be used when drywall sheets are installed with visible surface reveals as part of the finished product.

T F **21.** J- and U-trim are designed to be used when the edge of the drywall sheet is visible as a finished product.

T F **22.** Drywall screws are available in lengths from ¾″ to 4″.

T F **23.** Light-gauge metal studs are available in thicknesses of No. 25 and No. 20 gauge.

_____________ **24.** Hanger wire bundles contain precut wire in lengths of _________′ to 20′.

_____________ **25.** Paper joint tape is supplied in rolls _________′ long.

Prefinished Panel Trims

_____________ **1.** J-trim

_____________ **2.** Corner trim

_____________ **3.** Batten

_____________ **4.** F-corner trim

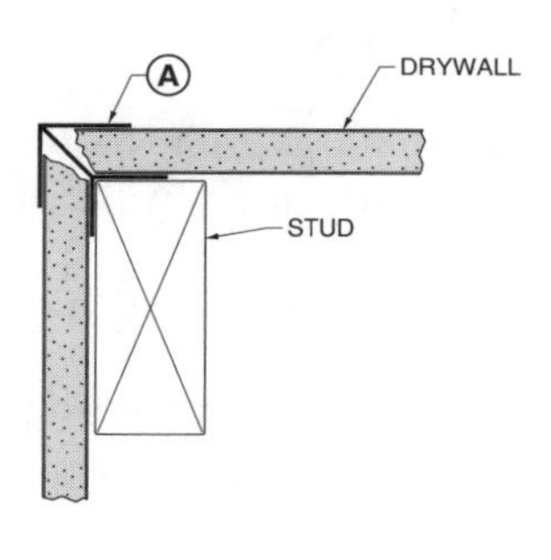

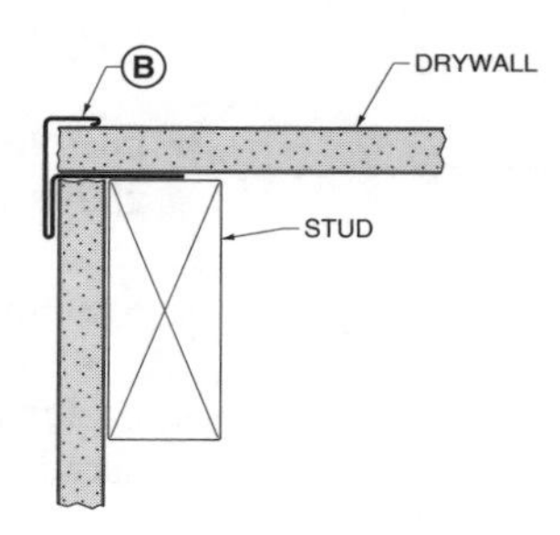

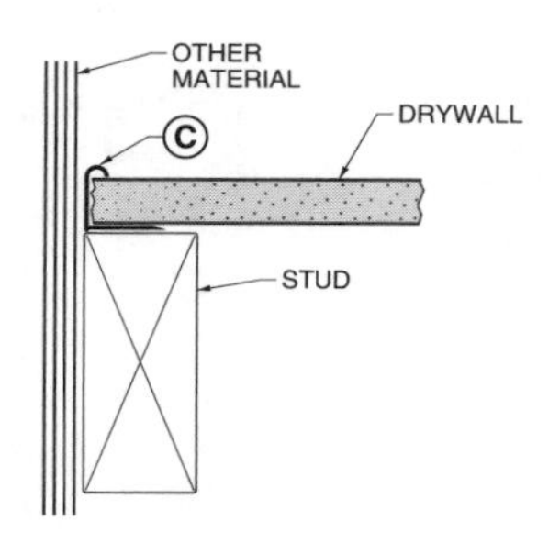

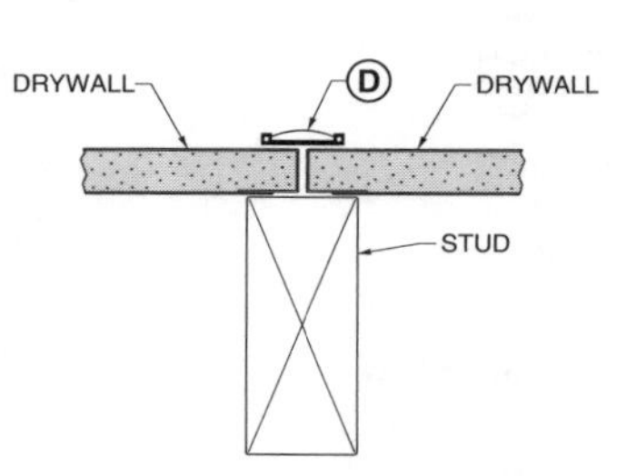

Light-Gauge Metal Studs

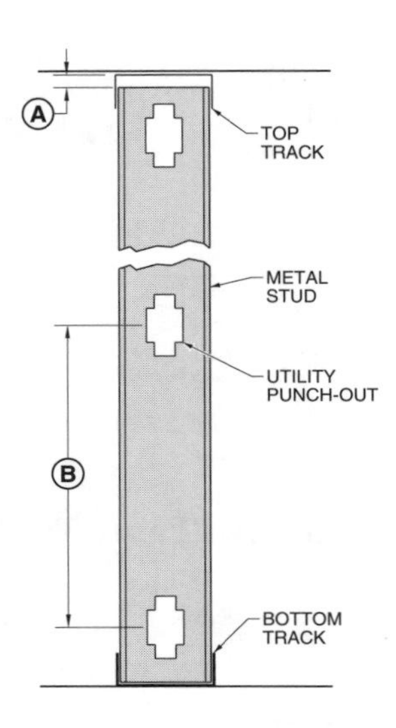

_____________ **1.** The recommended spacing at A is ________″.

_____________ **2.** Utility punch-outs at B are spaced ________″ OC.

Drywall Screws

_____________ **1.** Fastens single- and double-layer gypsum bases to steel studs, metal furring, and resilient channel

_____________ **2.** Fastens steel studs to steel runners

_____________ **3.** Fastens gypsum sheathing and gypsum base to No. 20 gauge or thicker steel studs in curtain wall assemblies

_____________ **4.** Fastens steel studs to runners, metal door frame

_____________ **5.** Fastens face-layer gypsum to base-layer gypsum in laminated partitions

Ⓐ S PAN HEAD

Ⓑ S-12 BUGLE HEAD

Ⓒ BUGLE HEAD

Ⓓ G BUGLE HEAD

Ⓔ S-12 PAN HEAD

3

Tools and Equipment

A tool is any device used to produce work. A hand tool is any tool powered by a human. A power tool is any tool powered by a source other than humans. Good safety practices should always be followed when using tools.

TOOLS

A *tool* is any device used to produce work. Tools are broadly classified as hand tools and power tools. Power tools include electrical-powered tools, powder-actuated tools, and pneumatic tools. Power tools may be portable or stationary. The major source of power for power tools is electricity.

Hand Tools

A *hand tool* is any tool powered by a human. Hand tools require no external power source to operate. Hand tools include tools used for measuring, striking, cutting, filing, tightening, leveling, etc.

Tape Measure. A *tape measure (steel tape)* is a hand-held measuring device with a retractable, graduated steel blade. Graduations are in the English system (feet, inches, and fractional parts of an inch), the SI metric system (meters, centimeters, etc.), or a combination of the two. Common blade lengths range from 12′ to 30′. Common blade widths range from ½″ to 1″.

Whichever length blade is preferred, the one selected must have a blade tip large enough to prevent it from slipping off the end of the metal framing components. The blade tip should be large enough to comfortably grip between the thumb and forefinger when marking drywall sheets and other materials with a pencil or when making cuts with a utility knife.

The tape measure is used for measuring, marking, and laying out drywall sheets and similar materials and for performing layout tasks on drywall sheets. See Figure 3-1. Metal framers working on commercial structures may select either a 16′ or 25′ tape measure for routine measuring and a 100′ tape measure for performing lay out. Metal framers working primarily on tenant partition work may use a tape measure with a 16′ blade. Drywall hangers who work mostly on residential construction hang 12′, 14′, and 16′ drywall sheets and may prefer a 25′ tape measure. Drywall workers who work primarily in commercial construction may prefer a 16′ tape measure since the drywall sheets used are seldom longer than 12′.

To assist in returning the tape blade into the case, one end of a steel spring is connected to the blade and the other end is connected to a center pivot located within the case. When the steel measuring tape is being used, the blade is pulled from the case and extended under tension from a spring. When returning the blade into the case, the drywall worker must be careful not to let the blade snap back into the case. Doing so causes the tip to strike the case opening with sufficient force to eventually break off the tip.

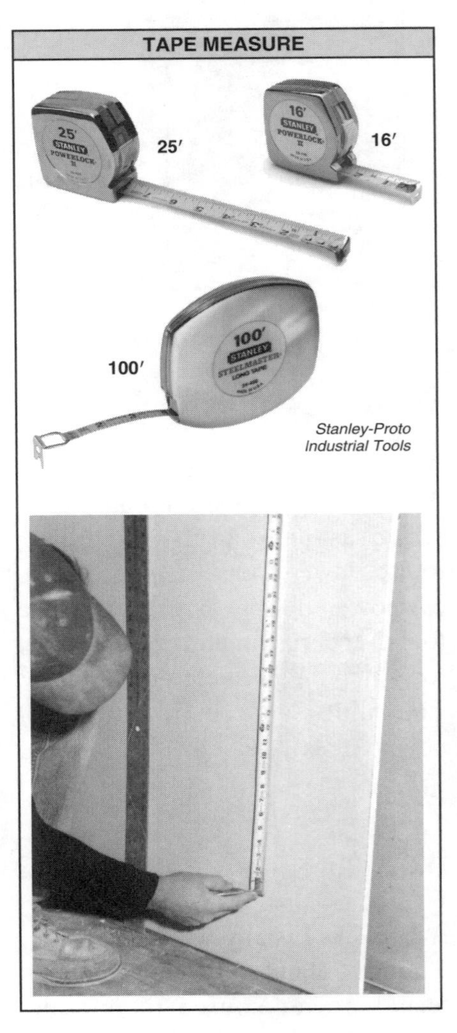

Figure 3-1. The tape measure is used for measuring, marking, and laying out drywall and similar materials.

To prolong the life of the tape measure and increase the efficiency of its use, the blade and internal parts must be kept dry and clean. Occasionally, a light film of oil is applied to the blade to ensure ease of operation.

When the tape measure is not being used, most right-handed drywall workers prefer to store it in the small pocket sewn into the nail bag directly above the nail pouch on their left side. See Figure 3-2. This location allows easy access to the tape measure with the left hand, which is used to hold it while performing measuring and cutting tasks. Most left-handed drywall workers place the tape measure in the same location on their right side.

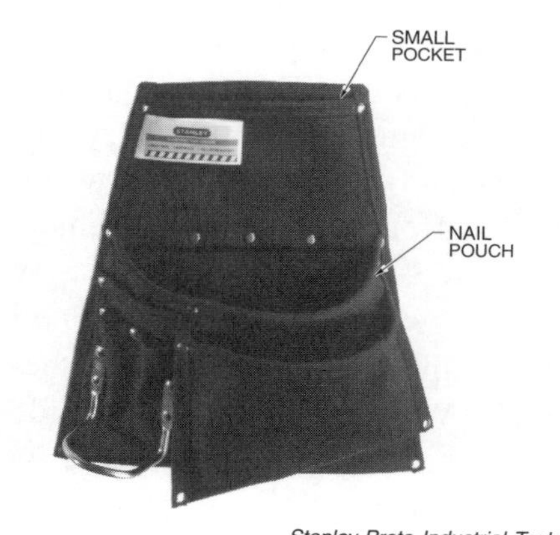

Stanley-Proto Industrial Tools

Figure 3-2. The tape measure is commonly carried in the small pocket of the nail pouch.

Red Devil, Inc.

Dust masks manufactured by Red Devil, Inc. provide protection against the inhalation of nontoxic dust, powders, and other common airborne irritants.

Steel Square. A *steel square* is an L-shaped layout tool with or without graduations. It is used to lay out 90° angles and mark small measurements. The preferred model has an 8″ tongue and a 12″ blade. See Figure 3-3. The tongue and blade of the steel square contain graduations in inches and fractional parts of an inch (1/8″, 1/4″, 3/8″, 1/2″, 5/8″, 3/4″, 7/8″). A carpenter's framing square, which has a larger tongue and blade, may also be used. Both are valuable tools for the metal framer and must be treated with care to avoid bending, nicking, or other damage.

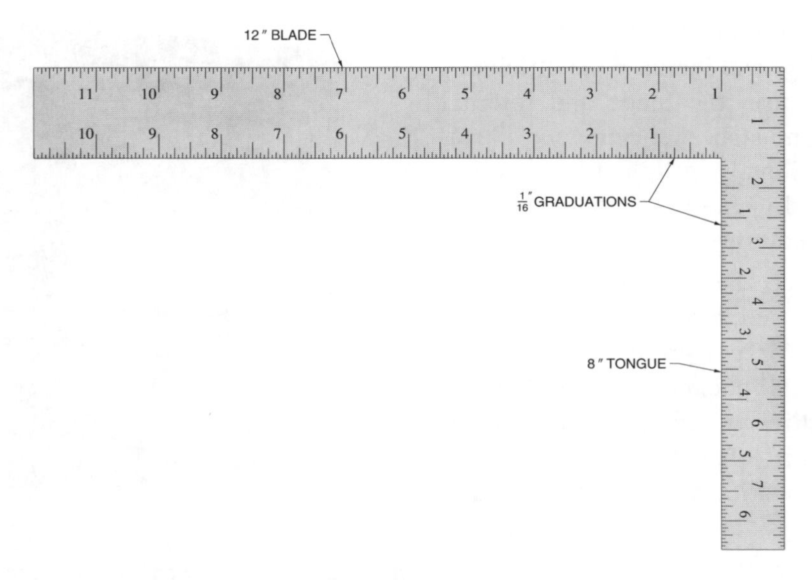

Figure 3-3. A steel square is an L-shaped layout tool with or without graduations.

T-Square. A *T-square (drywall T-square)* is a T-shaped layout tool with or without graduations. See Figure 3-4. The drywall T-square is constructed from two flat aluminum bars. It has a 24″ long top guide and a 48″ long blade joined to form the shape of the capital letter T.

Drywall T-squares may be purchased with or without graduations. Those with graduations are commonly divided into ⅛″ increments, which are printed along the edges of the top guide and blade. Those without graduations are generally less expensive. Both models are acceptable. Drywall T-squares must be handled with care since the alignment of the top guide with the blade must remain at 90°.

The drywall T-square is used as a guide for cutting or marking drywall sheets or other material. It is used as a guide for cutting drywall sheets and similar material when the location of the cut to be made is located at a distance not easily reached with a tape measure and knife (approximately 54″ and beyond). When it is used as a guide for cutting, the utility knife is placed on the material beside the blade. When it is used as a guide for marking, a line is drawn on the material beside the blade.

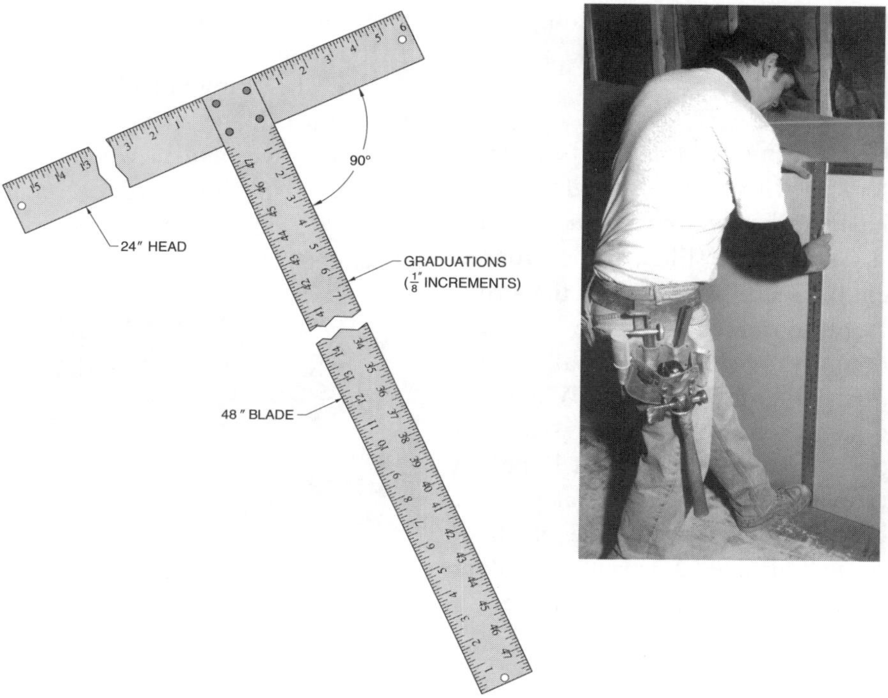

Figure 3-4. A T-square is a T-shaped layout tool with or without graduations.

When using the drywall T-square as a guide for making cuts, the following procedure should be followed:

1. Position the T-square with the top guide resting snugly against the edge of the material to be cut. Be sure that the intended location of the cut is immediately beside the edge of the blade.
2. Press the top guide firmly against the edge of the material with the left hand and press the toe or knee against the face of the blade at the bottom of the T-square (opposite for left-handed individuals).
3. Place the utility knife perpendicular to the edge of the blade and make the cut.

When making cuts on drywall, be sure that enough pressure has been placed on the knife blade to ensure that the full thickness of the paper has been completely cut.

Once the face paper is completely cut, the board can be broken at that location.

Chalk Line. A *chalk line* is a layout tool used to snap a straight line. It has a spring wound around a spool in a small case filled with powdered chalk. See Figure 3-5. As the string is pulled from the case, a small quantity of powdered chalk is deposited on the string. When the chalk-coated string is stretched tight against a surface, it is pulled outward and allowed to snap back, depositing a straight line of chalk on the surface. A variety of colors of powdered chalk is available.

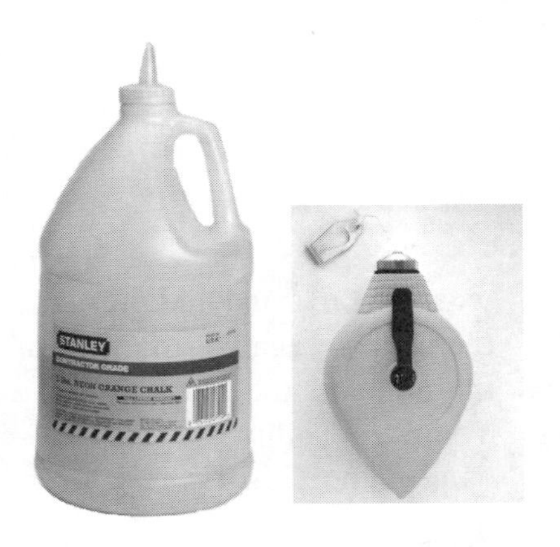

Stanley-Proto Industrial Tools

Figure 3-5. The chalk line is a layout tool used to snap straight lines.

Chalk lines are used to mark the layout of various framing components and the location of straight and angle cuts on drywall or similar material. They are also used when laying out and marking the location of floor and ceiling track. See Figure 3-6. The string must be kept dry at all times. If the string is allowed to become wet and is wound back into the case, the powdered chalk within the case becomes caked and unusable.

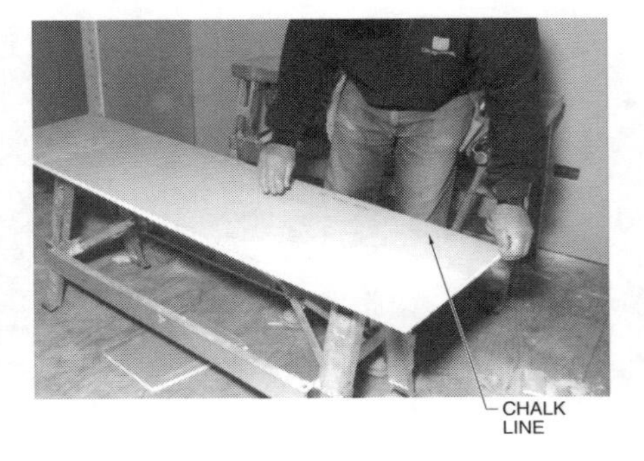

Figure 3-6. The chalk line can be used to lay out the location of floor and ceiling track.

Plumb Bob. A *plumb bob* is a layout tool used to establish a vertical line. See Figure 3-7. It is a cone-shaped metal weight fastened to a string. The force of gravity on the weight causes the string to hang in a vertical position. A plumb bob is used to transfer reference points that must be directly above or below each other. Plumb bobs are made from cone-shaped steel or brass and are available in sizes ranging from 8 oz to 32 oz. They have a point on the bottom end and a socket for attaching a string on the top end.

Stanley-Proto Industrial Tools

PLUMB BOBS		
Weight*	**Length****	**Diameter****
8	4½	1³⁄₁₆
12	5	1⅜
16	5¾	1½
24	6½	1¹¹⁄₁₆
32	7¼	1⅞

* in oz
** in in.

Figure 3-7. A plumb bob is a layout tool used to establish a vertical line.

When the layout marks have been established on the floor, it is necessary to transfer them to the structure overhead. Transferring to the desired location is accomplished by loosely holding the string of the plumb bob and allowing it to drop until it comes to rest slightly above the floor with the point directly above the reference mark. The overhead is marked at the point where the string touches. If the transfer is accurately made, the location of the upper mark is precisely above the bottom mark and the finished plane is plumb.

Spirit Level. A *spirit level* is a layout tool used to establish and check vertical and horizontal lines. See Figure 3-8. It is a sealed cylindrical tube (vial) with a slight curvature nearly filled with liquid, forming a bubble. The vial is enclosed in a wooden or aluminum frame of varying length. A 4′ aluminum level with a magnetic strip embedded along one side is helpful when plumbing door or window openings. The magnetic strip allows the spirit level to cling to the metal framing member, freeing both hands for anchoring the studs and track.

Aluminum levels must be handled with care to prevent being jarred out of alignment. They must be checked daily to ensure they are providing accurate readings. The alignment of a level may be checked by using the procedure:

1. Place level firmly against a vertical or horizontal surface.
2. Take a reading of the bubble position.
3. Turn level around so the opposite side is against the surface.
4. Take a reading of the bubble position.

If the position of the bubble remains at dead center during both positions, the level is properly adjusted. If there is a variation of the bubble position as the level is reversed, an adjustment is required.

Figure 3-8. A spirit level is a layout tool used to establish and check vertical and horizontal lines.

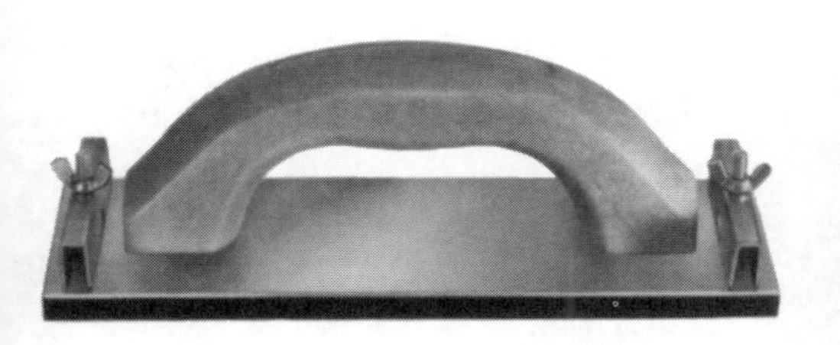

Hyde Tools

Drywall hand sanders from Hyde Tools have aluminum backed neoprene sanding pads and are used for hand sanding drywall joints.

Claw Hammer. A *claw hammer* is a striking tool with a slightly curved head used to drive nails and a slotted claw used to pull nails. See Figure 3-9. The claw hammer is available in a variety of sizes (12 oz, 16 oz, 22 oz, etc.), handle materials (metal, wood, fiberglass), and styles (straight claw or curved claw). Whichever hammer is selected, the edges of the claws and the face should be smooth for drywall work.

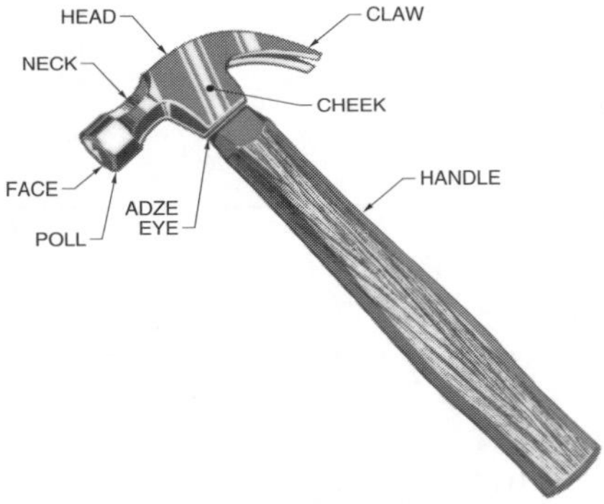

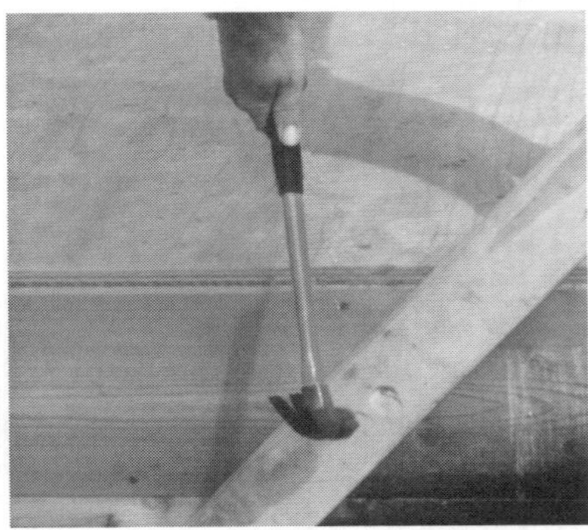

Figure 3-9. A claw hammer is a striking tool with a slightly curved head used to drive nails and a slotted claw used to pull nails.

Serrated-faced claw hammers are designed to prevent skidding off the nail head but must not be used by the metal framer. The face must be smooth to reduce the possibility of flying particles when striking metal framing components. If a claw hammer is purchased with sharp corners or edges, they may be filed smooth before use.

Drywall Hammer. A *drywall hammer* (drywall axe or hatchet) is a striking tool with a serrated face used to drive fasteners into drywall and leave a dimple. See Figure 3-10. Drywall hammers are available in a variety of sizes and styles. Drywall hangers choose a drywall hammer based on personal preference and one that best suits the type of work being performed. Whichever style is selected, the edges of the face should be smooth and the corners rounded, even though the face is serrated to prevent it from skidding off the nail head.

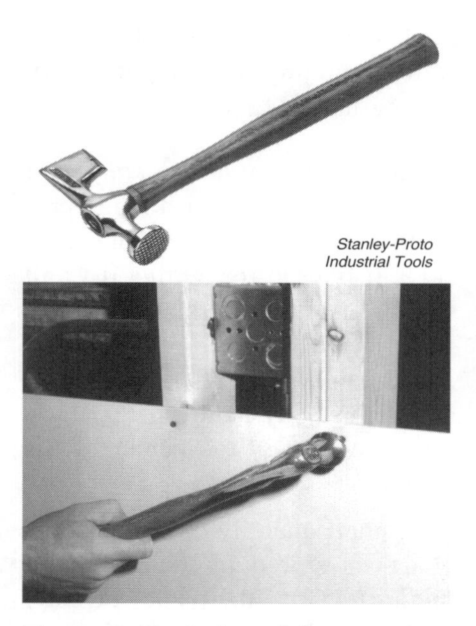

Figure 3-10. A drywall hammer is a striking tool with a serrated face used to drive fasteners into drywall sheets and leave a dimple.

A drywall hammer with a smooth convex face permits nails to be driven properly with good dimples and without cutting the face paper of the drywall sheet. Drywall hammers with sharp face edges will scar the drywall face paper and produce unacceptable dimples with cut edges.

The drywall hammer is used for a variety of tasks other than driving drywall nails. For example, the blade of the drywall hammer is used to pry drywall sheets into place and to mark the position of electrical boxes and other cutouts. Although the drywall hammer is made in the shape of a hatchet, it is not generally used for cutting or chopping.

When selecting a new drywall hammer, be sure to check for handle offset. If the drywall hammer has too little handle offset, the insufficient knuckle space between the handle and the face of the drywall may injure the knuckles when driving nails.

Utility Knife. A *utility knife* is a cutting tool with a short blade protruding from the handle. See Figure 3-11. It is used for cutting drywall sheets and similar material. The style of knife preferred by most drywall workers has a two-piece aluminum body held together by a screw located in the center of the handle.

A small quantity of replacement blades is stored in the handle of the knife. Once the blade currently being used becomes unusable, it is discarded and one of the replacement blades is installed in its place. Replacing the old blade with a new one is accomplished by removing the screw holding the two halves of the handle together and then carefully removing the replacement blade. When a utility knife is purchased, a small quantity of replacement blades is provided within the handle. Additional replacement blades are obtained separately from a material supply house or hardware store and are carried in the tool bucket or box.

The blade must be sharp when cutting drywall sheets or similar material. The potential for injury to the drywall worker is increased if the blade is dull. The blade may be sharpened several times using a sharpening stone before replacement is necessary.

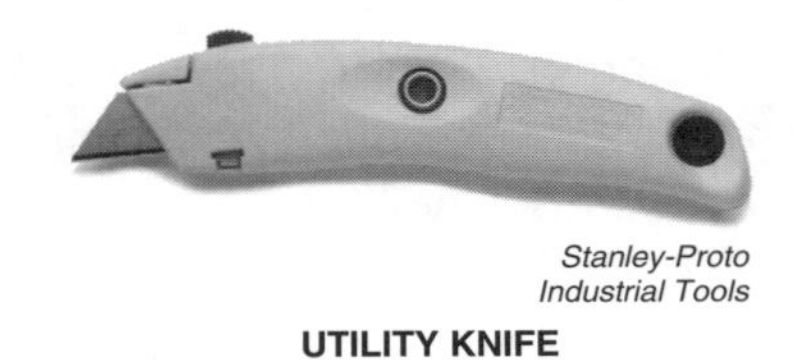

Stanley-Proto Industrial Tools

UTILITY KNIFE

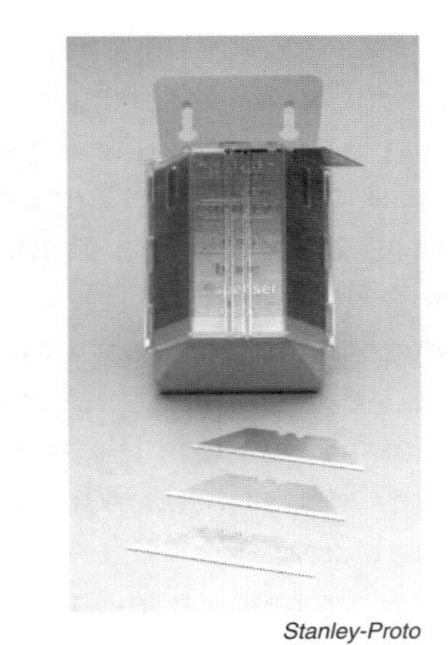

Stanley-Proto Industrial Tools

REPLACEMENT BLADES

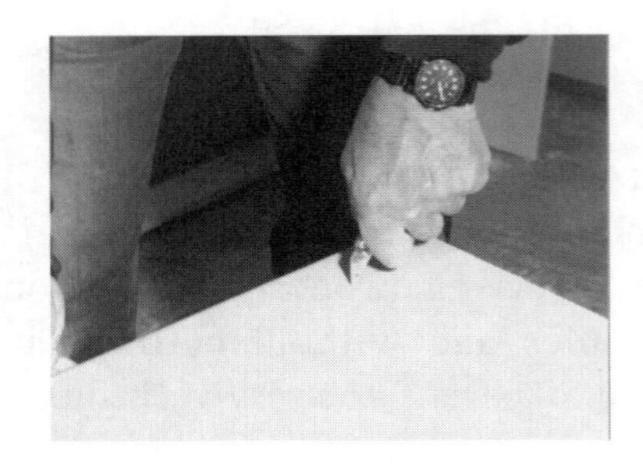

Figure 3-11. A utility knife is a cutting tool with a short blade protruding from the handle.

Cutting drywall sheets or similar material requires enough pressure on the knife blade to completely cut through the face paper of the drywall sheet and slightly score the core. A properly-executed cut allows the drywall sheet to be easily broken at the location of the cut. After the end of the drywall sheet has been broken to form a right angle, the utility knife is used to cut the back paper and the cut is complete. When it is not being used, most drywall workers carry the utility knife in the small knife pocket located at the front or rear of their tool pouch, where it is easily reached.

Rasp. A *rasp* is a coarse file used to shape wood, drywall, and other soft materials. See Figure 3-12. It is used to smooth the edges of the drywall sheet after it has been cut. This is especially necessary when cutting narrow rips for use on wraps and drops. Although many drywall installations do not require the cut edges to be rasped smooth, the rasp is a required tool and must be carried either in the tool pouch or tool bucket.

Rasps can be purchased from a tool supplier or hardware store, or one can be constructed by wrapping a piece of metal lath around a short scrap of 1″ × 4″ wood. A piece of metal lath is cut with sufficient length to cover the bottom and the ends of the 1″ × 4″. The metal lath is formed around the wood block and nailed at each side and end. Drywall cuts are rasped by holding the face of the rasp at a right angle to the face of the drywall and running it along the cut portion several times until the desired smoothness is achieved.

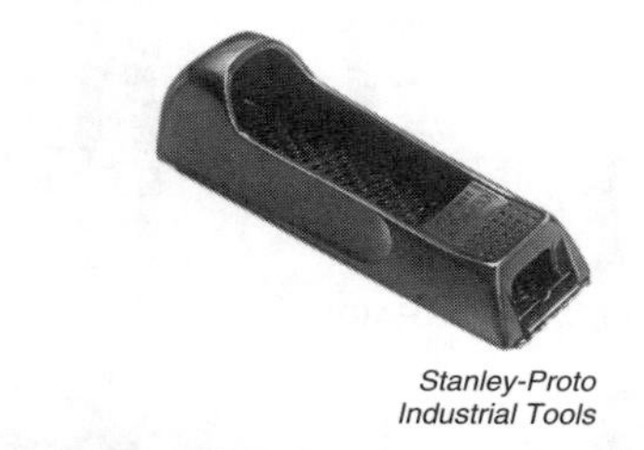

Stanley-Proto Industrial Tools

Figure 3-12. A rasp is a coarse file used to shape wood, drywall, and other soft materials.

Snips. *Snips* are a scissor-like hand tool used for cutting light-gauge metal and other materials. See Figure 3-13. Aviation snips are preferred by most metal framers. Snips are available in left-hand, right-hand, and straight-cut. Straight-cut snips are preferred for general metal cutting. Snips are used to cut light-gauge metal studs, metal stud track, furring channel, framing angle, resilient channel, etc. A short-nose model is also available for cutting heavy-gauge metal framing components.

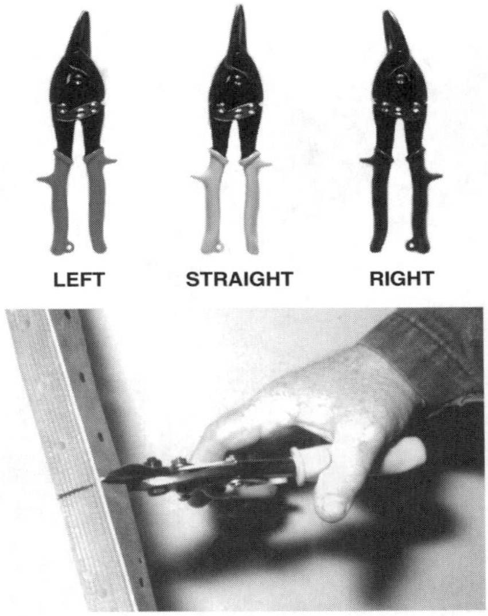

LEFT STRAIGHT RIGHT

Figure 3-13. Snips are a scissor-like hand tool used for cutting light-gauge metal and other materials.

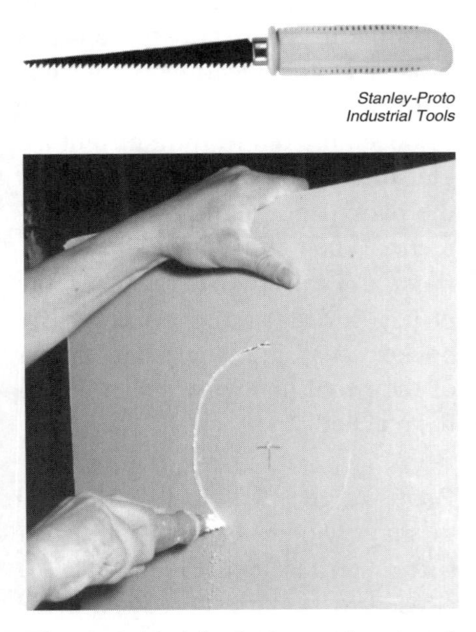

Figure 3-14. A keyhole saw is a handsaw with thin, tapered, interchangeable blades used for cutting curves and inside holes.

Keyhole Saw. A *keyhole saw* is a handsaw with thin, tapered, interchangeable blades used for cutting curves and inside holes. See Figure 3-14. It is used to saw holes for electrical boxes, large pipes, and other odd shapes. The point of the saw is sharp enough to pierce the drywall sheet when beginning the cut. This allows the drywall worker to press the saw against the face of drywall sheet and pierce the drywall with a twisting motion. Once the saw has pierced the drywall sheet, a sawing motion is used to complete the cut.

Hacksaw. A *hacksaw* is a metal-cutting handsaw with an adjustable steel frame for holding various lengths and types of blades. See Figure 3-15. It is used for a variety of metal-cutting operations. A hacksaw with a 12″ blade is preferred by most drywall workers because it provides a longer cut with each stroke. Hacksaw blades have a different number of teeth per inch depending on the material to be cut. The more teeth per inch, the finer the cut. The fewer teeth per inch, the coarser the cut. Fine-tooth blades cut metal framing members with less chatter than coarse-tooth blades and are preferred by drywall workers.

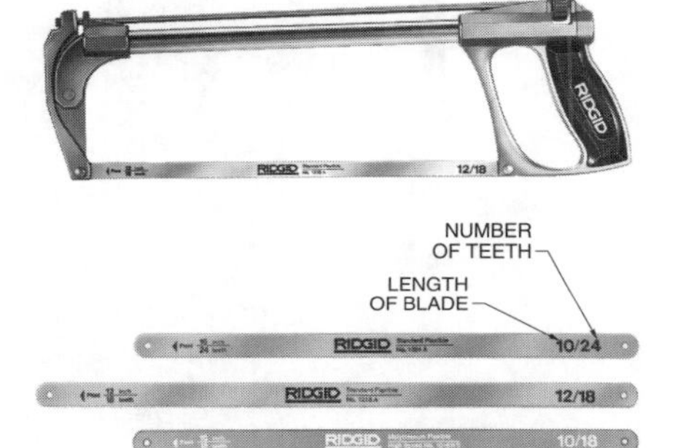

Ridge Tool Company

Figure 3-15. A hacksaw is a metal-cutting handsaw with an adjustable steel frame for holding various lengths and types of blades.

Circle Cutter. A *circle cutter* is a hand tool used to cut circles in thin wood or drywall sheets. See Figure 3-16. The circle cutter consists of a cutting wheel attached to a slide bar and a pivot point. The location of the cutting wheel is fully adjustable to accommodate circles ranging from 2″ to 16″ in diameter.

Cuts made with the circle cutter are begun by determining the diameter of the circle desired and setting the slide bar and pivot point to this dimension. The slide bar is locked in place by tightening the large thumb screw located above the pivot point. The pivot point is then pressed into the drywall sheet or other material at the center of the hole to be cut. Pressure is applied to both the pivot point and the cutting wheel, and the cutting wheel is rotated around the pivot point until the full thickness of the drywall face paper has been cut.

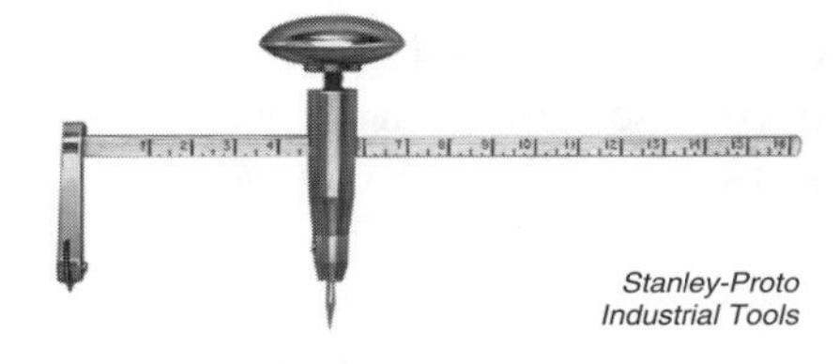

Stanley-Proto Industrial Tools

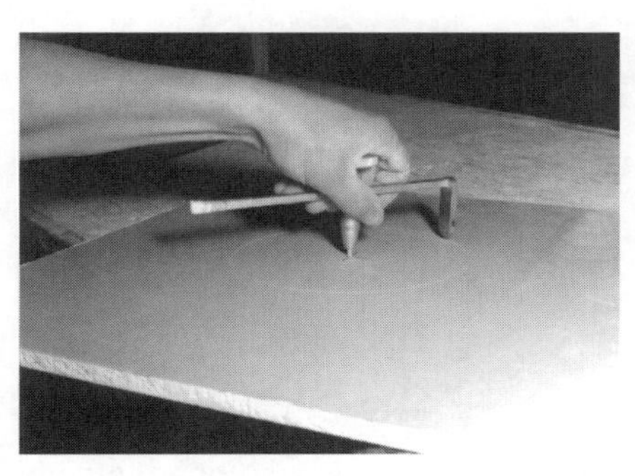

Figure 3-16. A circle cutter is a hand tool used to cut circles in thin wood or drywall sheets.

Sufficient pressure should be applied to completely cut through the drywall face paper and score the core. With the cut completed, the circle cutter is removed and the center portion is either sawed or knocked out with a drywall hammer. The debris is cleaned from within the circle and from the back of the drywall sheet. The uneven drywall remaining at the perimeter of the circle is removed with a utility knife.

Gypsum Board Stripper. A *gypsum board stripper* is a hand-held cutting tool that cuts both sides of a panel at the same time. See Figure 3-17. The handle guides adjustable cutting wheels along the edge or end of the panel. The maximum cutting depth of a gypsum board stripper is 4½″.

Wallboard Tool Co., Inc.

Figure 3-17. A gypsum board stripper is a hand-held cutting tool that cuts both sides of the panel at the same time.

Pipe-Hole Cutter. A *pipe-hole cutter* is a hand tool used to cut holes in drywall sheets that are too small to cut with a circle cutter. See Figure 3-18. The pipe-hole cutter is made from scrap materials usually found on the job. The pipe-hole cutter is a 4″ to 6″ piece of ¾″ thin-wall electrical conduit cut at approximately a 45° angle on one end.

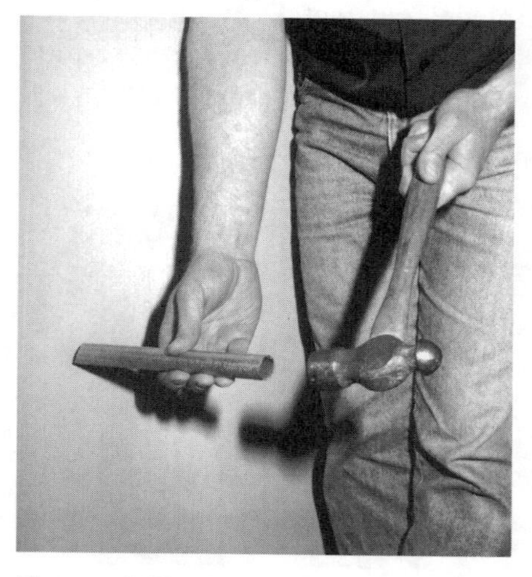

Figure 3-18. A pipe-hole cutter is a hand tool used to cut holes in drywall sheets that are too small to cut with a circle cutter.

The drywall worker holds the sharp end of the pipe-hole cutter against the face of the drywall sheet at the location marked and strikes it on the flat end with a sharp blow from the drywall hammer. This produces a clean hole for pipes that is small enough to be easily filled when the drywall sheet is taped and finished.

Phillips Head Screwdriver. A *Phillips head screwdriver* is a hand tool with a head designed to fit into a cross-slotted screw head for turning. See Figure 3-19. Although most framing and drywall screws are installed with an electric screwdriver, there are occasions when it is necessary to remove or tighten a framing screw or sink a drywall screw and no screwgun is handy. Most metal framers carry a Phillips head screwdriver in their tool pouch and another in their tool bucket or tool box.

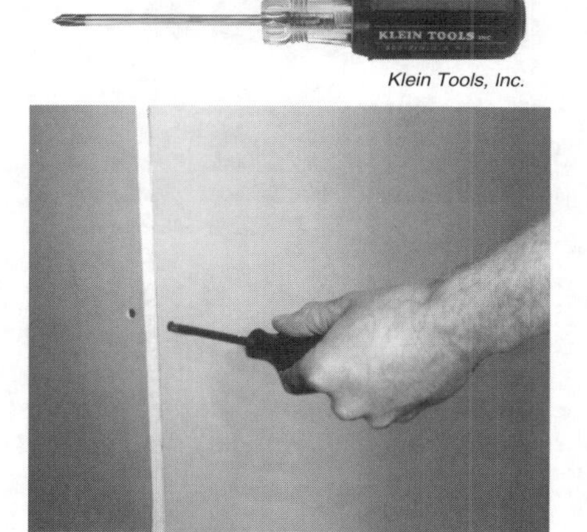

Klein Tools, Inc.

Figure 3-19. A Phillips head screwdriver is a hand tool with a head designed to fit into a slotted screw head for turning.

Locking "C" Clamps. *Locking "C" clamps* are C-shaped clamps used to apply pressure to material. See Figure 3-20. One end has a fixed jaw opposed by an adjustable jaw. A lever in the handle allows for quick release of the jaws. Locking "C" clamps provide a positive hold on framing components while assemblies are being completed. This allows the drywall worker the use of both hands for installing framing screws or making welded connections. Locking "C" clamps are available in several sizes. Drywall workers usually carry several pair in their tool bucket or tool box.

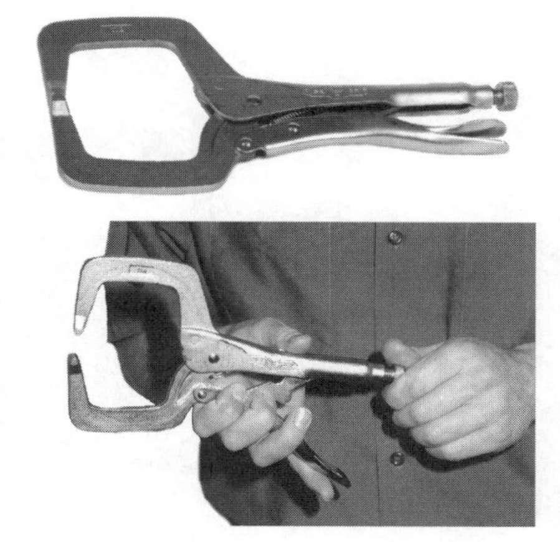

Figure 3-20. Locking "C" clamps are C-shaped clamps used to apply pressure to material.

Drywall Lifter. A *drywall lifter (kicker)* is a lifting device made of a short piece of metal that is tapered on one end and has a fulcrum on the bottom. See Figure 3-21. A *fulcrum* is the support about which a lever turns. The drywall lifter is used to lift drywall sheets into position prior to fastening.

The front end of the drywall lifter is placed on the floor against the bottom edge of the drywall sheet and kicked into position under the drywall. The rear end of the drywall lifter is pressed down with the foot, lifting the drywall sheet into position for fastening. The drywall lifter or kicker should be positioned as near the center of the sheet as possible so the weight is distributed evenly and one corner does not rise higher than the other corner.

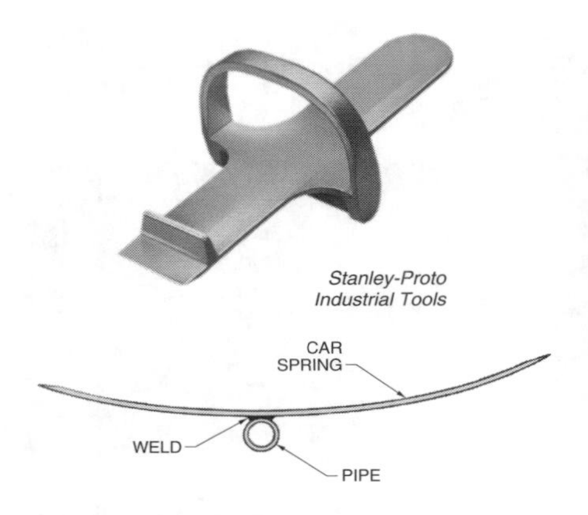

Figure 3-21. A drywall lifter (kicker) is a lifting device made of a short piece of metal that is tapered on one end and has a fulcrum on the bottom.

When bottom drywall sheets are applied last, they are placed next to the wall and lifted approximately ½″ until they are in firm contact with the previously installed top sheets. Drywall lifters can be purchased from a tool supplier or hardware store or they can be constructed by welding a short piece of pipe to the bottom of a short leaf spring from an automobile.

Aerial Platforms. An *aerial platform* is a lifting device for materials and workers. Self-propelled platforms have a standard lift capacity up to 2500 lb and can extend up to 70′. The wide range of working heights and lift capacities are useful in commercial and industrial applications. See Figure 3-22.

Figure 3-22. Aerial platforms provide a wide range of working heights and lift capacities.

Taping (Mud) Pan. A *taping (mud) pan* is a joint compound carrier for hand finishing. See Figure 3-23. It is available in a wide range of sizes, with or without a knife-cleaning blade. Taping pans are made of plastic, stainless steel, galvanized steel, and tinplate. The standard taping pan is 4″ wide, 13″ long, and 3¾″ deep, with slightly sloping sides. Some drywall workers use a plaster hawk to hold the taping compound in place of a taping pan.

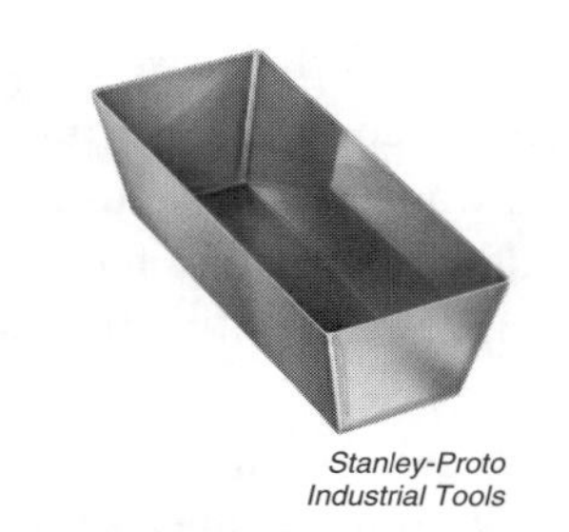

Stanley-Proto Industrial Tools

Figure 3-23. A taping (mud) pan is a joint compound carrier for hand finishing.

Taping Knife. A *taping knife* is a hand tool used to wipe down the tape after it has been applied. See Figure 3-24. The 6″ or 7″ taping knife is the taper's universal tool. It is used to wipe down the tape, apply the first coat of joint compound to metal or plastic drywall trim, and spot nails or screws. Taping knives that have solid metal handle butts can be used to pound in protruding drywall nails. Taping knives with 8″, 10″, and 12″ blades are used for applying second and third coats of joint compound as well as joint topping compound. Taping knives with 2″ and 4″ blades are used for applying joint compound in tight spots such as corners and areas around pipes and other mechanical fixtures.

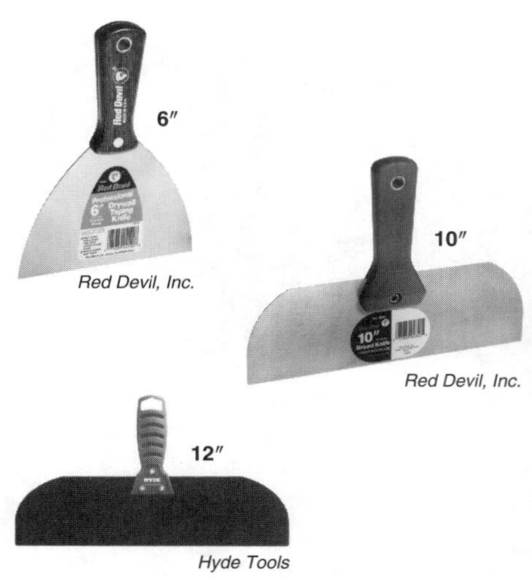

Figure 3-24. A taping knife is a hand tool used to wipe down the tape after it has been applied.

Joint Compound Mixer. A *joint compound mixer (stomper)* is a long-handled hand tool used for mixing powdered and premixed joint compound. See Figure 3-25. The stomper is worked up and down in the bucket to thoroughly mix the joint compound. An electric drill motor equipped with a special paddle is also available for mixing joint compound.

Figure 3-25. A joint compound mixer is a long-handled hand tool used for mixing powdered and premixed joint compound.

Pole Sander. A *pole sander* is a sanding tool with sandpaper attached to a swivel head on a pole. See Figure 3-26. It is used to sand dried joint compound between coats and upon final touch-up. The pole sander consists of a metal head with clamps that firmly hold a sheet of sandpaper attached to a wooden or metal handle with a swivel connection. Various grades of precut sheets of sandpaper are used by the drywall worker, depending on the operation being performed. Coarse grades are used for rough sanding and fine grades are used for final touch-up. A protective dust mask should be worn at all times when sanding drywall.

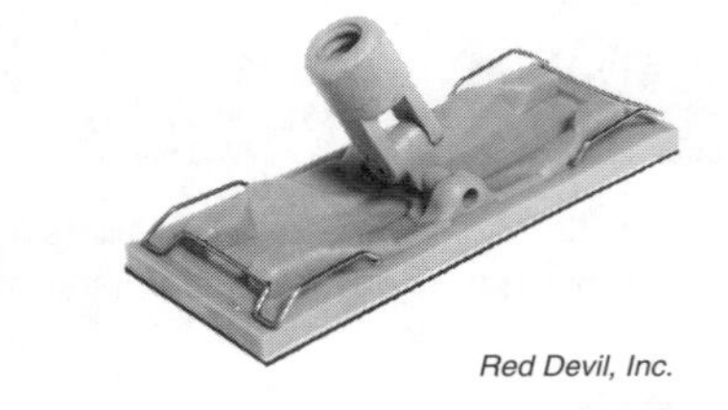

Red Devil, Inc.

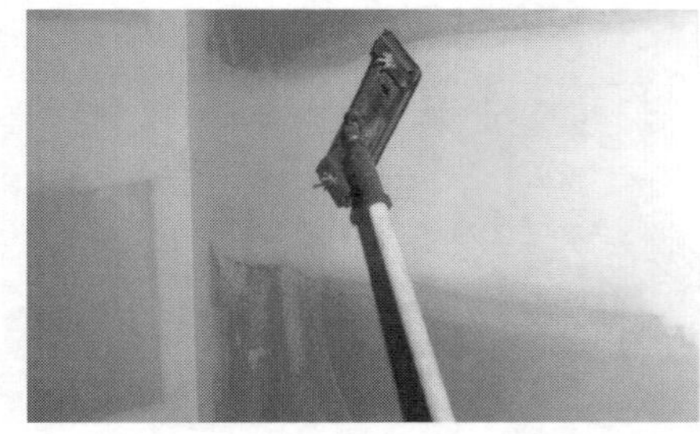

Figure 3-26. A pole sander is a sanding tool with sandpaper attached to a swivel head on a pole.

Mechanical Taping Tools. The mechanical taping tools used for applying drywall tape and successive coats of joint compound speed the application and covering of the drywall tape. See Figure 3-27. A complete set of mechanical drywall tools consists of:

- Taping machine (Bazooka®)
- Angle roller
- Angle skimmer
- Angle skimmer handle
- Angle box
- Angle plow
- 8″ skim box
- 10″ skim box
- 12″ skim box
- Skim box handle
- Handle extension
- Nail spotter
- Pump
- Pump goose neck
- Skim box filler attachment

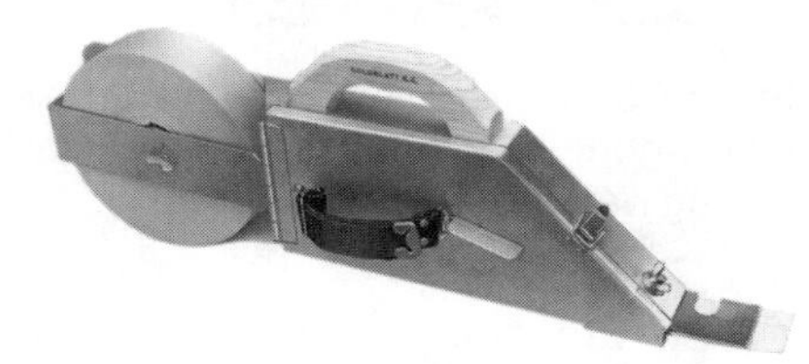

Stanley-Proto Industrial Tools

Figure 3-27. The mechanical taping tools used for applying drywall tape and successive coats of joint compound speed the application and covering of the drywall tape.

An alternative taping machine (banjo) is also available. All mechanical drywall taping tools are designed to be easily disassembled for cleaning and servicing. Mechanical drywall taping tools must be thoroughly and properly cleaned on a daily basis to ensure trouble-free operation.

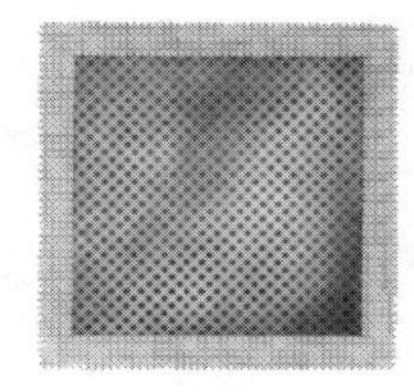

Hyde Tools

Self-adhesive fiberglass mesh wall patches from Hyde Tools eliminate the need for pre-plastering for repair of drywall, plaster, wood, and most other materials.

Power Tools

A *power tool* is any tool powered by a source other than humans. Power tools may be portable or stationary. Portable power tools are generally lighter and smaller than stationary power tools. Portable power tools may be easily transported to and used on the job site.

Electrical-powered tools have a grounding plug or an insulated body. See Figure 3-28. They are the most common power tools. Those with a metal body have a three-prong electrical cord which contains a grounding connection. Those with a plastic or other non-conductive material body are insulated and have a two-prong electrical cord. These insulated power tools are designed to shield the worker from contact with electrical power.

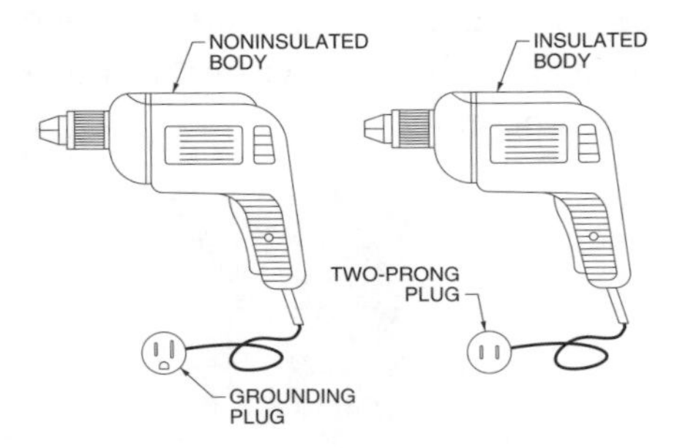

Figure 3-28. Electrical-powered tools have a grounding plug or an insulated body.

Electric Screwdriver. An *electric screwdriver (screwgun)* is a power tool used for driving various types of screws. See Figure 3-29. The screwgun consists of an electric drill motor with a clutch-controlled shaft, a magnetic screw-holding tip, and an adjustable nosepiece in place of the drill chuck. The clutch-controlled shaft permits the tip to stop rotating when the correct torque has been reached.

The depth of the drywall screw is controlled by the adjustable nosepiece. The entire assembly has been constructed to allow the rotation of the tip to stop once the framing screw has been driven tight or once the desired drywall screw depth has been reached. The magnetic screw-holding tip holds the screw until it is driven into place when driving framing screws or when the screw has penetrated the surface of the drywall sheet when driving drywall screws. The screwgun shaft assembly is spring-loaded and rotates only when the screw placed on the magnetic tip of the screwgun is pressed against a surface.

Figure 3-29. An electric screwdriver (screwgun) is a power tool used for driving various types of screws.

The electric screwgun has a switch-lock button which is depressed to lock the switch in the ON position. For normal drywall hanging applications, the screwgun is turned ON and the switch lock is depressed. This keeps the motor of the screwgun running when numerous screws are being driven at one time. Otherwise, the trigger must be held in the ON position or repeatedly pressed to start the motor. Although the motor may be running continuously, because the magnetic tip does not rotate until the tip is pressed against a surface, placing a new screw on the tip is always accomplished while the tip is stopped. It is essential that the adjustable nosepiece of the screwgun be properly adjusted so the drywall screws are driven to the correct depth for applying the joint compound. See Figure 3-30.

Hilti, Inc.

Figure 3-30. The screwgun is adjusted so the drywall screws are driven to the correct depth.

When the electric screwgun is used exclusively for metal framing operations, the nosepiece is usually removed and stored. Removing the nosepiece of the screwgun exposes the magnetic shaft. This allows the shaft and magnetic tip better access to tight spots when performing various framing tasks.

The nosepiece must be in place when the electric screwgun is used to screw drywall sheets to metal or wood framing and other materials. Always test the depth setting before beginning to install drywall. Begin the process by driving a few test screws into the drywall sheet to determine whether the proper depth setting has been made. The proper depth setting for drywall screws is just slightly below the surface of the face paper on the drywall sheet.

If the screwgun tip, which is held in place by the magnetic shaft, breaks or becomes unusable, it must be replaced. This is accomplished by removing and replacing the tip.

1. Unplug the screwgun from the power source.
2. Remove the nosepiece from the screwgun.
3. Remove the damaged tip by pulling it forward.
4. Clean any debris from the magnetic shaft receptor.
5. Insert the new tip into the magnetic shaft.
6. Adjust the nosepiece for the proper screw depth.

Manufacturer's operating instructions and safety precautions must be followed when using a screwgun. Safety glasses must be worn when using an electric screwgun. Ear protection is also recommended.

½″ Electric Drill Motor. A *½″ electric drill motor* is a power tool used for drilling holes. See Figure 3-31. It has a chuck with a ½″ capacity. The ½″ electric drill motor is used for drilling holes in concrete and other materials. This power tool is also used for mixing drywall joint compound when equipped with a special mixing paddle.

Ridge Tool Company

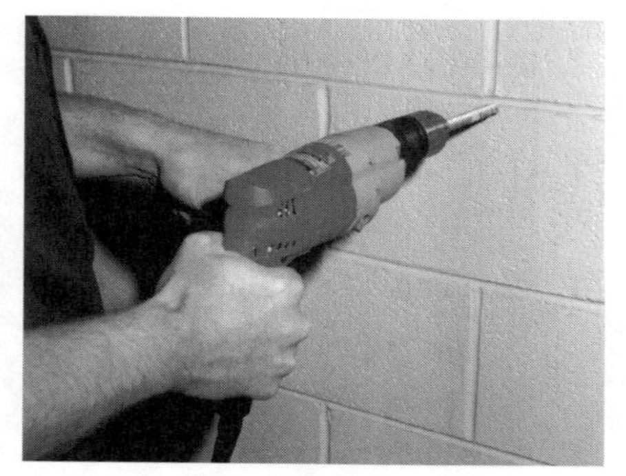

Figure 3-31. A ½″ electric drill motor is a power tool used for drilling holes.

The extension cord must be of an adequate gauge wire to ensure that the tool does not become overheated. When using an electric drill motor, the designed rpm must be maintained. The tool must not be overloaded too much, or it slows below its operating speed. This causes the tool to overheat and may cause the motor windings to burn out completely. Eye protection must be worn when drilling holes with an electric drill motor. Ear protection is required if the work being performed generates excessive noise.

Drywall Cutout Tool. A *drywall cutout tool* is a hand-held electric cutout tool used to make irregular cuts and holes in panels, gypsum board, etc. for electrical boxes, duct openings, etc. See Figure 3-32. The drywall cutout tool is lightweight and is used with one hand. Its high rpm allows it to cut through panel products quickly. A variety of bits are available for cutting different materials.

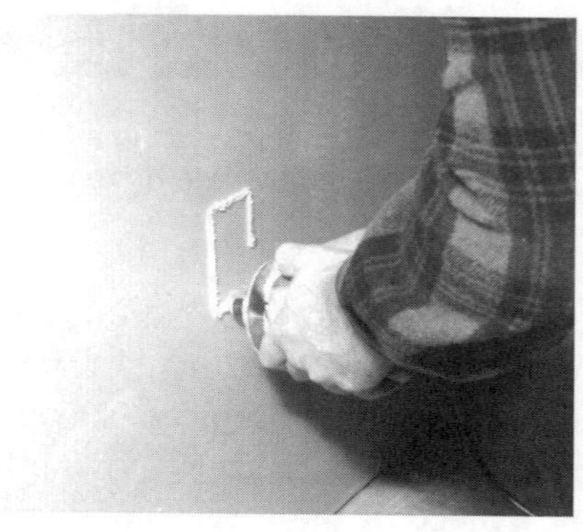

Figure 3-32. A drywall cutout tool is a hand-held electric cutout tool used to make irregular cuts and holes in panels, gypsum board, etc. for electrical boxes, duct openings, etc.

Cut-Off Saw. A *cut-off saw* is a light-duty portable electric saw used for cutting material to length. See Figure 3-33. It consists of an electric motor containing a shaft which holds a Carborundum saw blade. The saw blade is mounted on a swing arm over an adjustable table. It is covered by a protective guard which retracts as the cut is made. Safety glasses or a face shield and ear protection must be worn at all times while using a cut-off saw. Usually, this saw is taken to the place where the material to be cut is located, rather than moving the material to the saw location. This increases efficiency and reduces the number of times the material must be handled.

Porter-Cable Corp.

Figure 3-33. A cut-off saw is a light-duty portable electric saw used for cutting material to length.

The cutting process should not be rushed when the cut-off saw is used for cutting either light- or heavy-gauge metal framing components. The saw blade cuts more efficiently when allowed to run at normal speed. Rushing the cut slows the speed of the blade and places added resistance on the motor, causing it to build up heat. Continued use under these circumstances may cause the windings of the motor to overheat and burn out.

A heavy-duty stationary electric cut-off saw is a high-capacity saw, usually mounted on wheels, designed to be set up at a convenient location on the job site. Usually this saw is placed adjacent to the framing material stockpile to reduce the distance the material must be moved to the saw. This may be less efficient than using a portable cut-off saw, however, the higher capacity of the stationary cut-off saw may be necessary due to the type and size of the material being cut.

Portable Arc Welders. An *arc welder* is a shielded metal arc welding (SMAW) machine in which the arc is shielded by the decomposition of the electrode covering. See Figure 3-34. SMAW equipment includes the arc welder, cables, electrodes, electrode holder, and grounding clamp.

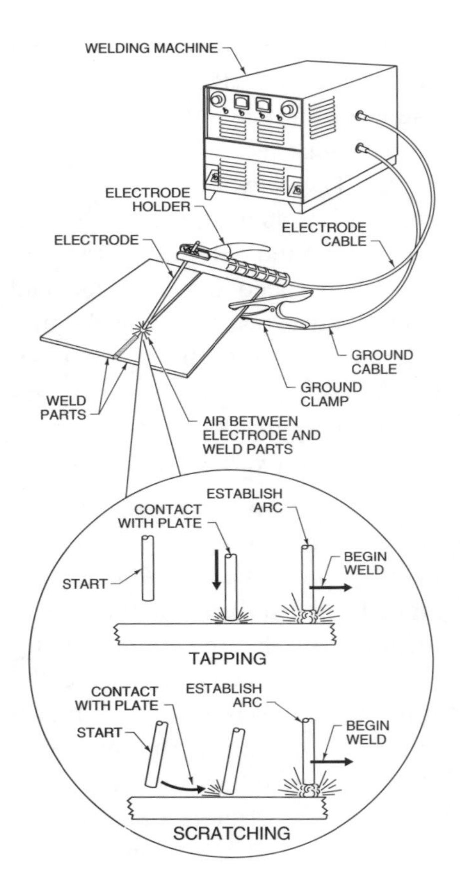

Figure 3-34. An arc welder is a shielded metal arc welding (SMAW) machine in which the arc is shielded by the decomposition of the electrode covering.

An arc is struck to begin the SMAW process. The electrode touches the metal being welded to complete the electrical circuit. The electrode is withdrawn while maintaining the arc between the electrode and the work. Electricity flows from the welding machine, through the electrode cable, through the electrode holder, through air, to the weld parts, through the ground clamp, through the ground cable, and back to the welding machine.

Two types of portable arc welding machines are used by metal stud framers. One type uses a gasoline engine to drive a direct current generator. The other type (referred to as a buzz box) uses a transformer to convert standard 115 V or 230 V electrical power into the low-voltage, high-amperage electrical current needed to perform the welding operation.

Arc welding rods (electrodes) are made from heavy steel wire which has been coated with a special insulating material to prevent the rod from becoming a filament. As the welding rod is lightly raked or struck across the surface being welded, a flame or arc is created. Once the arc is properly started, the tip of the welding rod is raised slightly above the work and oscillated with a slight side-to-side or circular movement.

Wire-feed welding machines are used in addition to the standard gasoline and transformer arc welding machines, which use a coated steel rod. See Figure 3-35. The wire-feed welding machine does not use a conventional welding rod holder (stinger). Instead, the positive lead is constructed from a hollow tube and the wire is continuously fed through the tube as needed. Superior welds are possible when welding light-gauge framing components using a wire-feed welding machine.

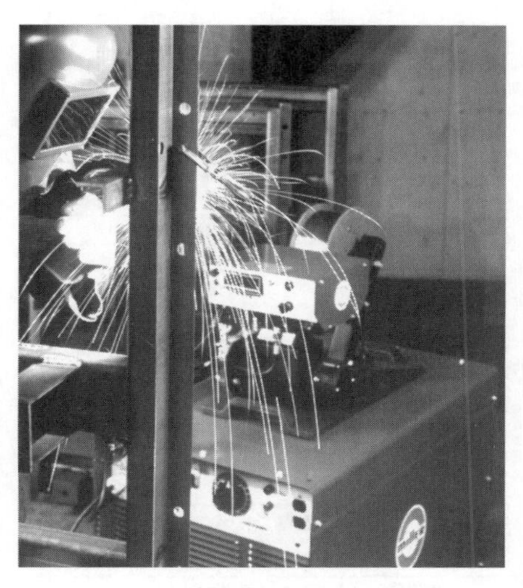

Miller Electric Manufacturing Company

Figure 3-35. Wire-feed welding machines are used for welding light-gauge framing components.

The construction plans and specifications must be consulted prior to performing any welding operations. They provide the necessary information regarding welding rod type and the width, length, and spacing of each weld. Strict compliance with these requirements is mandatory. Many times the structural integrity of the building is dependent on the framing members being attached to each other and to the building superstructure with suitable welds. An inspection of all welds used for structural assemblies is required by the construction specifications or the local building code. These inspections must be made prior to the installation of wallcovering materials.

A welding face shield with tinted safety glass and a long-sleeved leather welding jacket must be worn while welding. This prevents serious flash burn damage to the eyes, face, arms, and chest resulting from the heat and sparks which fly in all directions as the welds are made.

Laser Level. A *laser level* is a leveling device in which a concentrated beam of light is projected horizontally or vertically from the source and used as a reference for leveling or verifying horizontal or vertical alignment. See Figure 3-36. A laser level has a rotating head which transmits a thin beam of red light. It can be mounted in either the horizontal or vertical position.

When used for framing interior walls or partitions, the laser level is placed on its side to transfer the layout from the floor to the walls and overhead structure. This allows the framer to position the top track adjacent to the light beam as it is attached to the overhead structure. When the laser level is used for suspended ceiling installations, it is mounted horizontally so the beam provides a level reference point throughout the area where the work is being performed. This allows the carrier channels to be positioned with extreme accuracy, ensuring a flat and level finished ceiling.

Figure 3-36. A laser level is a leveling device in which a concentrated beam of light is projected horizontally or vertically from the source and used as a reference for leveling or verifying horizontal or vertical alignment.

Electrical Cords. Electrical extension cords are essential to the use of electrical-powered tools and must be handled properly. Serious electrical shocks can result from worn or frayed electrical cords. The grounding prong on all grounded electrical cord plugs must be intact and should never be snipped off to accommodate plugging in a two-prong receptacle. Electrical extension cords are available in different wire sizes (gauge) and plug configurations. See Figure 3-37.

THREE-WIRE EXTENSION CORD RATINGS*

Extension Cord Length**	Ampere Rating on Tool Nameplate						
	0 – 6	8	10	12	14	16	20
	Wire Size***						
25	18	18	18	16	14	14	12
50	18	18	18	16	14	14	12
100	18	16	16	16	14	14	12
150	16	16	14	14	12	12	12

* 125 V
** in ft
*** AWG

WIRING DEVICES

2-POLE, 3-WIRE NONLOCKING

WIRING DIAGRAM	NEMA ANSI	RECEPTACLE CONFIGURATION	RATING
	5-15 C73.11		15 A 125 V
	5-20 C73.12		20 A 125 V
	5-30 C73.45		30 A 125 V
	5-50 C73.46		50 A 125 V

2-POLE, 3-WIRE LOCKING

WIRING DIAGRAM	NEMA ANSI	RECEPTACLE CONFIGURATION	RATING
	ML2 C73.44		15 A 125 V
	L5-15 C73.42		15 A 125 V
	L5-20 C73.72		20 A 125 V

Figure 3-37. Electrical extension cords are available in different wire sizes (gauge) and plug configurations.

The most common extension cords used to supply electrical power to portable electrical-powered tools contain three separate wires enclosed in an insulated jacket. The color of the insulation on each wire (black, white, or green) indicates the purpose of the wire. Black insulation is used for the hot wire, which is connected directly to the source of electrical power. White insulation is used for the neutral wire, which completes the flow of electrical current when the tool is in use. Green insulation is used to indicate the ground wire, which is connected to the ground bus at the electrical source and provides a direct path to ground in case of a short circuit. The ground wire is connected to the frame of the electrical-powered tool.

The correct polarity must be maintained when connecting electrical extension cords to the power source. *Polarity* is the particular state of an object, either positive or negative, which refers to the two electrical poles, north and south. To ensure polarity, the wires are connected black-to-black, white-to-white, green-to-green, etc.

When selecting an extension cord, be sure that the wire size (gauge) that is stamped on the side of the cord jacket is adequate for the electrical requirements of the tool being used. If the gauge is too small for the tool, the extension cord wire heats up and the efficiency of the tool is reduced. If the length of the extension cord exceeds the design limits, a voltage drop occurs and causes the motor of the tool to heat up. The correct gauge of the extension cord required for a tool is determined by the ampere rating shown on the nameplate of the tool and the distance from the power source. If, for example, the ampere rating of the tool being used is 14 A and the length of the extension cord is 100′, the gauge of the cord must be at least No. 14.

Powder-Actuated Tools

A *powder-actuated tool* is a device that drives fasteners by means of an explosive charge. Powder-actuated tools are used for anchoring framing components to concrete, masonry, and steel. Powder-actuated tools have a firing mechanism similar to a gun. Blank cartridges contain powder which, when activated, generates the force which propels a fastener or a piston down the barrel of the tool. The amount of force generated by the cartridge gives the fastener sufficient velocity to drive it into the base material or causes a piston to drive the fastener with sufficient velocity to drive it into the base material. Powder-actuated tools are available as direct-acting and indirect-acting types. See Figure 3-38. *Note:* Always follow manufacturer's recommendations, operating procedures, and safety precautions when operating any powder-actuated tool. Always start with the lowest possible power level and increase until a proper fastening is made.

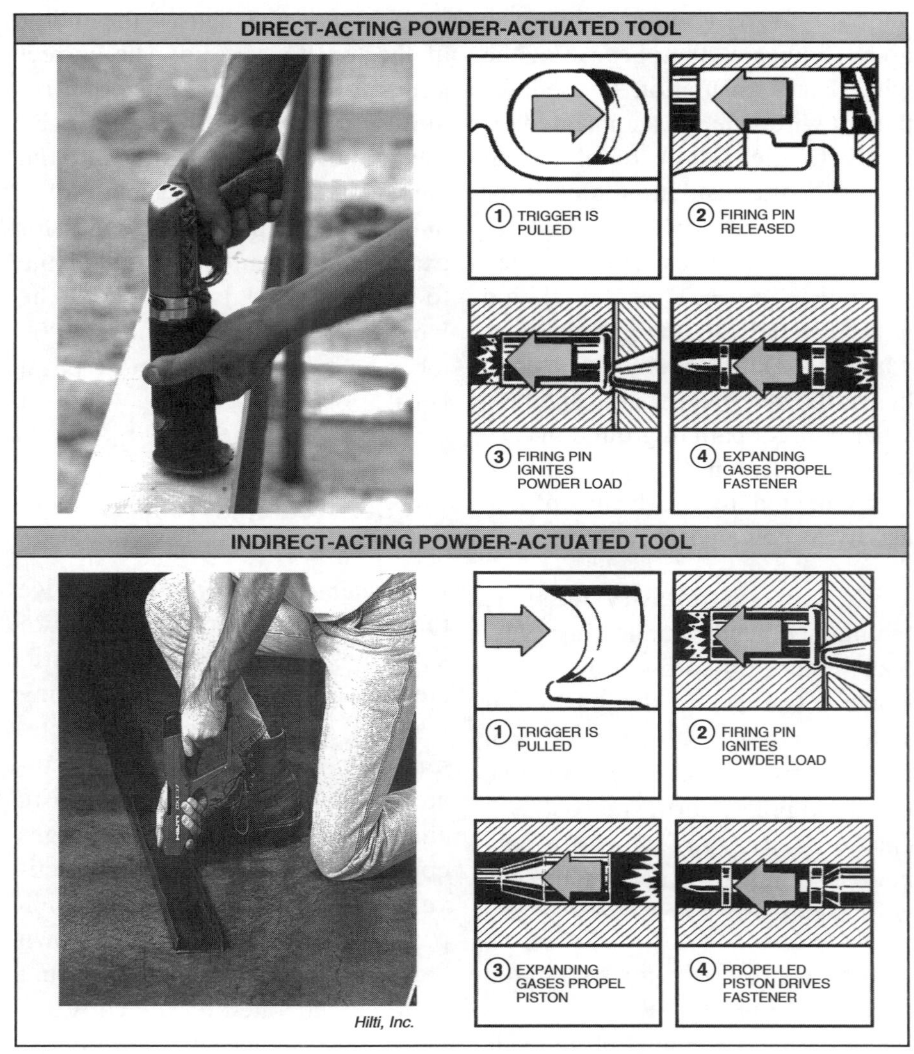

Figure 3-38. A powder-actuated tool is a device that drives fasteners by means of an explosive charge.

In a direct-acting powder-actuated tool, the force created by the exploding gunpowder in the cartridge propels the fastener down the barrel of the tool and into the work. To operate a direct-acting powder-actuated tool, apply the procedure:

1. Turn the back part of the tool counterclockwise one quarter turn to the open position and break open the tool.
2. Remove the cartridge holder.
3. Expel the spent cartridge and store the cartridge holder on the pin provided near the opening.
4. Insert a fastener into the chamber with the point facing toward the front of the tool.
5. Insert a cartridge into the cartridge holder.
6. Place the cartridge holder and cartridge in the chamber on top of the fastener and press the fastener into the barrel.
7. Return the tool to the closed position by turning clockwise one quarter turn.
8. Press the tool against the work and pull the trigger to fire the cartridge.

In an indirect-acting powder-actuated tool, the force created by the exploding gunpowder in the cartridge propels a piston which propels a fastener into the work. To operate an indirect-acting powder-actuated tool, apply the procedure:

1. Open the chamber by dropping the front of the tool downward with a flick of the wrist. The barrel slides forward to eject the spent cartridge and open the chamber.
2. Insert a fastener into the chamber with the point facing toward the front of the tool and push down the barrel.
3. Insert a cartridge into the chamber.
4. Close the chamber.
5. Press the tool against the work and pull the trigger to fire the cartridge.

Hyde Tools

A pail opener is a device used to open plastic joint compound pails and buckets without damaging the lid seal.

A wide variety of fasteners are available for both types of powder-actuated tools. The variety of fastener types, lengths, and diameters is required because of the wide range of framing components that must be fastened and the different types of base material. Construction plans and specifications or the local building code requirements provide the necessary information regarding fastener length, diameter, and spacing.

Hyde Tools

Texture brushes manufactured by Hyde Tools are used to produce textured surfaces on walls and/or ceilings.

Cartridges for indirect-acting (low-velocity) powder-actuated tools are available in four power levels. The power levels of cartridges for low-velocity powder-actuated tools are indicated by a color-coded system. Power level 1 (light) is gray, level 2 (medium) is brown, level 3 (strong) is green, and level 4 (heavy) is yellow. Low-velocity cartridges are short with crimped front ends. Use only low-velocity cartridges in low-velocity powder-actuated tools. Powder-actuated tools may use single cartridges or a clip of 10 cartridges. The clip of 10 cartridges automatically places a cartridge in the chamber each time the tool is closed.

Cartridges for direct-acting (standard-velocity) powder-actuated tools are available in 12 power levels. The six lightest levels are indicated by color-coded brass cartridges. The color codes for the first four standard-velocity cartridges are the same as those used for low-velocity cartridges. Standard-velocity cartridges are available in two additional power levels. Power level 5 is red and power level 6 is purple. The next six power levels are available in color-coded nickel cartridges. The color codes for these are the same as the first six power levels. Blank cartridges are available in .22, .25, and .27 caliber. The caliber of the cartridges are matched to the tool used.

The wide range of power levels and the ability to adjust the expansion of the explosive gases provide infinite adjustment for any application. When power requirements fall between those provided by the powder charges, the power adjustment is made by pushing the fastener or piston part way down the barrel. This expands the chamber and reduces the force of the powder charge. All powder-actuated tools are supplied with an adjustment rod that is used to push the fastener or piston the required distance into the barrel. Standard-velocity cartridges are longer than low-velocity cartridges. Use only standard- velocity cartridges in standard-velocity powder-actuated tools.

Fastener Length. The proper length fastener must be selected when using powder-actuated tools. See Figure 3-39. For concrete, the fastener should penetrate through the thickness of the material being attached and penetrate 1″ to 1¼″ into the concrete. For steel, the proper length fastener is determined by adding the thickness of the material being attached and the thickness of the steel. The point of the fastener should penetrate through the steel slightly.

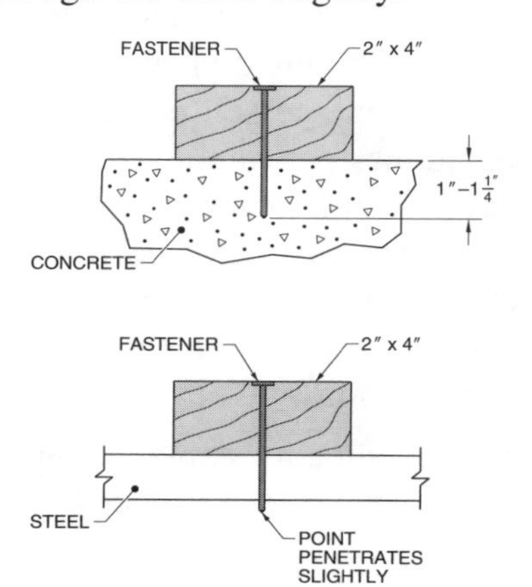

Figure 3-39. A wide variety of anchor pins are available for both types of powder-actuated tools.

Powder-Actuated Tool Safety. The American National Standards Institute (ANSI) publication, A10.3-1985, *Safety Requirements for Powder Actuated Fastening Systems*, details safe procedures to be followed when using powder-actuated tools. General safety rules for powder-actuated tools include:

- Never attempt to use a powder-actuated tool until properly trained and licensed.
- Follow manufacturer's instructions.
- Double-check to ensure that all walls are made of concrete if the surface to be fastened into has a plaster or other finish. This can be done by driving a nail through the surface covering. Do not fire into a concrete wall that is less than 2″ thick.
- Wear safety goggles to protect eyes from flying particles.
- Wear ear protection. Prolonged use of a powder-actuated tool can cause hearing damage.
- Do not stand next to or work in the immediate area of a person operating the tool. A fastener shot into concrete may glance off a piece of reinforcing steel in the concrete and fly to one side.
- Do not leave a loaded tool lying around. Do not load the tool unless it will be fired immediately.

- Post warning signs where powder-actuated tools are in use.

Pneumatic Tools

A *pneumatic tool* is a tool powered by compressed air. See Figure 3-40. The compressed air is regulated to control the pounds per square inch (psi). Compressors are powered by gasoline engines or electric motors. Common pneumatic tools used by drywall workers include staplers and T-nailers.

Figure 3-40. A pneumatic tool is a tool powered by compressed air.

When using a compressor powered by a gasoline engine, the engine oil must be checked regularly. Fuel must not be allowed to spill on a hot manifold as it may be ignited by the heat. When using a compressor powered by an electric motor, be sure that the gauge of the power cord is adequate for the size of the motor.

All pneumatic tools require occasional cleaning and oiling. Air hoses must be maintained in good condition and the fittings must be securely tightened. Air compressors must be serviced at regular intervals to ensure that they are in good working order. The air tank must be emptied daily to remove the moisture that has accumulated from condensation. The filters should be cleaned regularly and the belts must be kept in good working condition.

Stapler. A *stapler* is a pneumatic tool which drives staples into drywall and wood. See Figure 3-41. The staplers are widely used because of their labor-saving features. A drywall hanger or lather can drive an average of 10 to 15 conventional drywall or lathing nails per minute. Using a pneumatic stapler, the same person can drive 50 to 60 staples per minute. This represents a 400% increase in productivity without a similar decrease in quality and reduces the time required to complete the nailing portion of the job by 75%.

The pneumatic stapler is operated by the following procedure:

1. Press the end of the stapler firmly against the work.
2. Pull the trigger to drive the staple.

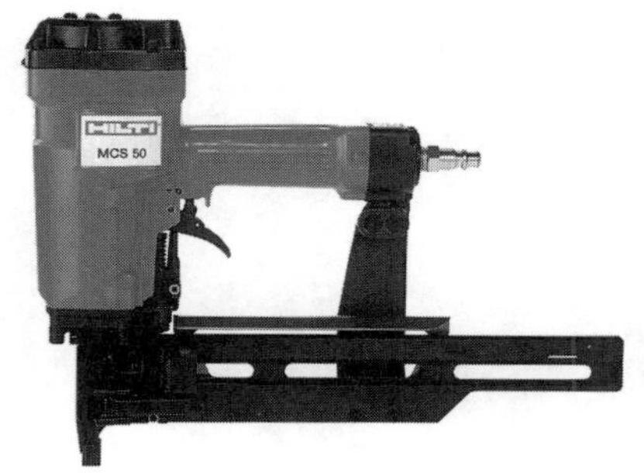

Hilti, Inc.

Figure 3-41. A stapler is a pneumatic tool which drives staples into drywall and wood.

Compressed air forces a piston to drive the staple into the work. Staples are expelled from pneumatic tools at a high rate of speed and could cause a serious injury if accidentally fired while not pressed against an acceptable surface. To reduce the potential for injury, a safety feature is built into pneumatic tools which prevents accidental firing. The tools do not function unless pressed against a firm surface. Pneumatic staplers must be properly adjusted to ensure that the staples are driven to the correct depth.

Pneumatic staplers are often used to fasten the first layer or layers of drywall to the wood framing members when performing multiple-layer drywall applications. The final layer is either applied with adhesive and a few conventional drywall nails or is nailed off completely with conventional drywall nails driven by hand. Eye and ear protection must be worn while operating the pneumatic stapler.

T-Nailers. A *T-nailer* is a pneumatic tool which drives hardened T-nails into concrete or masonry surfaces. See Figure 3-42. T-nailers must be properly adjusted to accommodate the density of the concrete or masonry into which the T-nail is being driven. They are used to attach metal stud tracks to concrete floors and metal studs and metal furring channels to masonry walls. This method for anchoring metal stud tracks, metal studs, and metal furring channels is often preferred because the cost of installing these framing members with T-nails is approximately 25% of the cost of using powder-actuated pins and blank cartridges. T-nailers are operated in the same way as pneumatic staplers. Eye and ear protection must be worn while operating T-nailers. Gloves are recommended.

Hilti, Inc.

Figure 3-42. A T-nailer is a pneumatic tool which drives hardened T-nails into concrete or masonry surfaces.

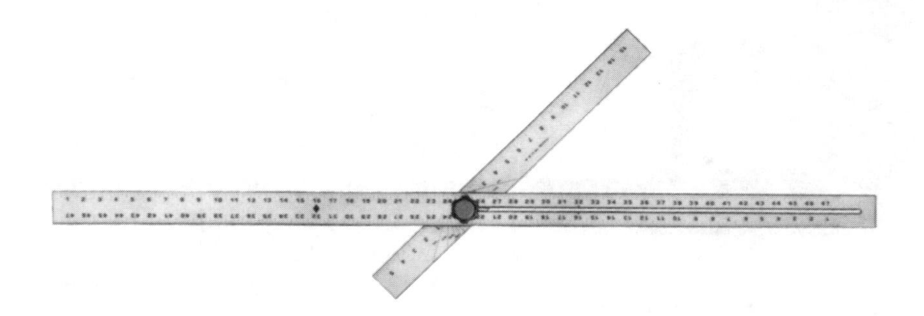

Hyde Tools

The adjustable T-square from Hyde Tools has a head that adjusts to any angle and locks in position at 90°. It is designed for drywall cutting and plywood, sheet metal, and general layout work.

GENERAL SAFETY

The most common injuries which occur to drywall hangers are cuts resulting from the improper handling of the drywall knife. Cutting drywall sheets is a relatively simple procedure. However, many injuries due to cuts from the drywall knife occur during this simple procedure. The best safety rule to follow while using the drywall knife is to position hands and fingers where they will not be injured if a slip of the knife occurs.

Drywall hangers receive serious injuries by misuse of the drywall hammer. When driving drywall nails, the drywall hangers must concentrate on the work being performed and be especially careful when working in tight or awkward locations.

Drywall tapers receive injuries resulting from falls due to slipping on wet joint compound carelessly dropped on the floor. They also receive cuts from the sharp edges of taping knives and paper cuts from drywall tape. Both drywall hangers and tapers suffer from strained back muscles due to carrying excessive loads.

All metal framing members and metal trim components have sharp edges and ends. Accidentally rubbing against these sharp edges and ends can result in serious cuts. When using aviation snips for cutting light-gauge materials, be sure that the cut portions are not allowed to come in contact with unprotected hands or fingers.

Safety Rules for Operating Powder-Actuated Tools

- Safety glasses (or a face shield) and ear protection must be worn at all times when using powder-actuated tools. Gloves are recommended.
- Powder-actuated tools must not be pointed at anything that is not to be fastened.
- Always be sure that the type of material the pin will be fired into is solid.
- Powder-actuated tools have a built-in safety feature which prevents the firing pin from being released unless the front of the tool is pressed firmly against the work.
- Never attempt to attach anything to hollow or brittle material.
- Do not fire the tool in the direction of other workers.
- Anchor pins must not be driven close to the edge of concrete or steel members.
- Most states require the operators of powder-actuated tools to have a license. Do not operate these tools without checking with the AHJ.

Safety Rules for Operating Pneumatic Tools

- Safety glasses (or a face shield) and ear protection must be worn at all times when using pneumatic tools. Gloves are recommended.
- Always be certain of the type of material into which the nails or staples will be driven.
- Do not drive nails or staples into hollow or thin material.
- Do not fire the tool in the direction of other workers.
- Keep both hands away from the area being nailed or stapled.

Scaffold and Ladder Safety. Rolling scaffolds must be properly constructed and braced and must be fitted with hand rails and toe boards. A rolling scaffold containing drywall sheets should not be moved until the drywall sheets have been installed. Scaffolds with drywall sheets become top-heavy and obstructions on the floor may cause them to tip over. Accidents of this type could result in serious injury or death. Ladders used for installing metal framing or drywall sheets must be long enough so that the drywall hanger does not have to stand on the top four steps or rungs. The ladder must also be sturdy and have antiskid pads on the bottom. Metal framers, drywall hangers, and drywall tapers must not reach more than 18″ beyond the ladder when installing framing members or drywall sheets or when performing taping operations. If the work is beyond the easy reach of the person performing the work, the ladder must be moved. It takes only a few moments to move a ladder and prevent a serious injury.

Personal Protective Equipment. Tradesworkers use personal protective equipment to prevent injury. See Figure 3-43. All personal protective equipment must meet OSHA Standard 29CFR1900-1999, applicable ANSI standards, and other safety mandates. Protective helmets or hard hats protect electrical workers from impact, falling and flying objects, and electrical shock.

Clothing made of durable material such as denim provides protection from contact with sharp objects and rotating equipment. Safety glasses, respirators, ear plugs, and gloves are used based on the work task being performed. For example, gloves made from rubber may be used to provide maximum insulation from electrical shock hazards. Safety shoes with steel toes and thick soles provide protection from sharp falling objects. Insulated rubber boots and rubber mats provide insulation to prevent electrical shock.

Figure 3-43. Tradesworkers use personal protective equipment to prevent injury.

Tools and Equipment

Trade Competency Test

Name ______________________ Date ______________

Tools and Equipment

______ **1.** A(n) ______ tool is any tool powered by a human.

______ **2.** Common blade widths of tape measures range from ______.

A. ¼″ – ¾″
B. ½″ – ¾″
C. ½″ – 1″
D. neither A, B, nor C

______ **3.** Common blade lengths of tape measures range from ______.

A. 10′ – 25′
B. 12′ – 25′
C. 10′ – 30′
D. 12′ – 30′

T F **4.** A steel square always has graduations on the blade.

T F **5.** The drywall T-square may be used for marking lines and as a guide for cutting.

______ **6.** A drywall hammer has a ______ face.

A. concave
B. convex
C. flat
D. A or C

______ **7.** A(n) ______ knife is a cutting tool with a short blade protruding from the handle.

______ **8.** A(n) ______ is a coarse file used to shape wood, drywall, and other soft materials.

______ **9.** The circle cutter can cut holes from 2″ to ______″ in diameter.

______ **10.** The maximum cutting depth of a gypsum board stripper is ______″.

__________ **11.** A joint compound mixer is also known as a(n) _________.

__________ **12.** A(n) _________ tool is any tool powered by a source other than humans.

__________ **13.** A(n) _________ saw is a light-duty portable electric saw used for cutting material to length.

T F **14.** A laser level can be mounted in either a horizontal or vertical position.

T F **15.** A powder-actuated tool is a device that drives fasteners by means of an explosive charge.

T F **16.** Bruises are the most common injury which occur to drywall hangers.

__________ **17.** The drywall T-square has a(n) _________″ blade.

__________ **18.** A(n) _________ is a layout tool used to snap a straight line.

__________ **19.** A(n) _________ is a layout tool used to establish a vertical line.

__________ **20.** A(n) _________ is a layout tool used to establish a level line.

__________ **21.** _________ are a scissor-like hand tool used for cutting light-gauge metal and other materials.

__________ **22.** A(n) _________ saw is a hand saw with thin, tapered, interchangeable blades used for cutting curves and inside holes.

__________ **23.** A(n) _________ is a metal-cutting handsaw with an adjustable steel frame for holding the blade.

__________ **24.** A(n) _________ is also known as a kicker.

__________ **25.** A(n) _________ tool is powered by compressed air.

Standard-Velocity Power Levels

______________	**1.** Power level 1	**A.** Red
______________	**2.** Power level 2	**B.** Green
______________	**3.** Power level 3	**C.** Purple
______________	**4.** Power level 4	**D.** Yellow

Arc Welder

______________ **1.** Ground cable

______________ **2.** Weld parts

______________ **3.** Welding machine

______________ **4.** Electrode cable

______________ **5.** Electrode

______________ **6.** Electrode holder

______________ **7.** Ground clamp

______________ **8.** Air between electrode and weld parts

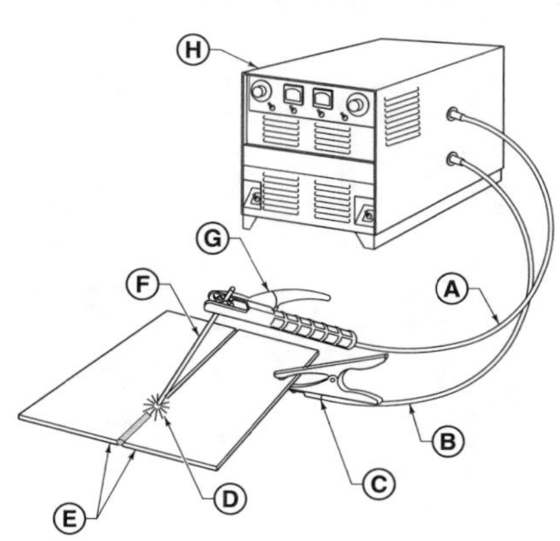

Extension Cords

_______________ **1.** A 25′ extension cord used with a portable power saw having a nameplate rating of 16 A shall have No. _________ wire.

_______________ **2.** A drill motor drawing _________ A can operate on a 50′ No. 16 extension cord.

_______________ **3.** The maximum length No. 18 extension cord for a 5 A load is _________′.

_______________ **4.** A 150′ extension cord with No. 16 wire can be used with tools having ampere ratings up to _________ A.

_______________ **5.** The maximum length No. 16 extension cord for a 10 A load is _________′.

THREE-WIRE EXTENSION CORD RATINGS*							
Extension Cord Length**	Ampere Rating on Tool Nameplate						
	0 – 6	8	10	12	14	16	20
	Wire Size***						
25	18	18	18	16	14	14	12
50	18	18	18	16	14	14	12
100	18	16	16	16	14	14	12
150	16	16	14	14	12	12	12

* 125 V
** in ft
*** AWG

4 Drywall Handling and Installation

Drywall sheets must be handled carefully to avoid damage. They should always be stored flat in a dry area. Drywall may be installed with nails, screws, or adhesives. A variety of trim is available for finishing drywall edges.

HANDLING DRYWALL

Proper handling of all gypsum board materials is essential to obtaining an acceptable finished product. Damage to drywall sheets occurring during the delivery and stocking process makes hanging more difficult and time-consuming. It also increases the time and effort required for the tapers during the taping and finishing operations. This procedure should be followed when handling drywall sheets:

- Handle all drywall sheets as though they are fragile.
- Be sure that the face side and edges of each sheet are not dented, scarred, or otherwise damaged.
- Do not open the two-sheet bundles until ready to hang.

The drywall hanger's involvement in the stocking process is usually limited. Drywall sheets must be unloaded from the delivery truck, moved into the building, and placed at predetermined locations convenient for the drywall hangers. This service is usually provided by the material supplier or an independent stocking agent. Occasionally, drywall hangers and apprentices or trainees are asked to assist with stocking drywall, but normally this work is performed by construction laborers.

The nature of the job may not allow the stacks of drywall sheets to be scattered throughout the building. When this condition exists, apprentices may move the drywall sheets from a central stocking area to the specific project location as they are needed by the drywall hangers.

In addition to following the proper lifting and carrying practices, these rules must be followed when stocking drywall:

- Never overload a drywall cart in order to reduce the number of trips.
- Do not roll drywall carts over extension cords or air hoses.
- Stack drywall sheets flat on boards on the floor.
- Do not stock drywall sheets by leaning them against a wall or other vertical surface.
- Use caution when working near forklifts or other mechanical lifts.
- Be aware of overhead power lines or other obstructions.
- Use only nylon web slings designed and rated for hoisting drywall.
- Inspect slings daily to be sure that they are not cut or excessively worn.
- Landing platforms must be properly anchored to the structure and must have safety rails and toeboards.

When drywall sheets are stored, even if for only short periods, they must be kept flat and dry. See Figure 4-1. If the supports placed under drywall stacks are spaced too far apart, the drywall sheets sag between the supports and become rippled or bowed. This greatly increases the effort required to hang and finish the drywall sheets. Stacks of drywall sheets must be located away from high-traffic areas. This reduces the possibility of damage from moving carts and other sources. Drywall stacks must not be used as workbenches.

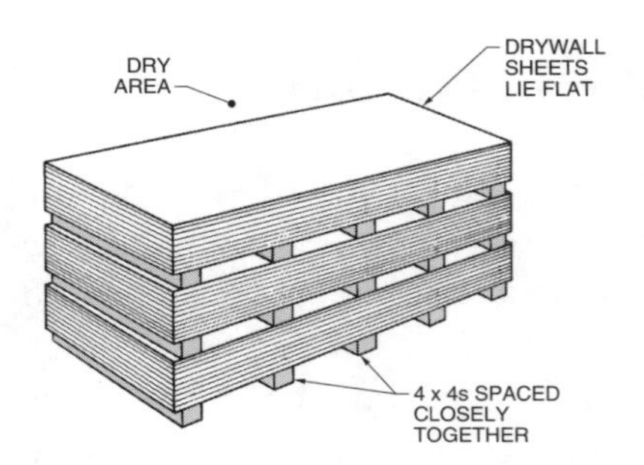

Figure 4-1. Drywall sheets must be kept flat and dry when stored.

Be sure drywall stacks are located within the building in such a manner that they do not overload the building structure. Drywall stacks must be placed directly over main structural beams or bearing walls. A representative of the general contractor should be contacted before the stocking operation begins.

Porter-Cable Corp.

The Model 6640 Drywall Driver from Porter-Cable Corp. is a variable speed reversing driver that has a 1⅝″ drywall screw capacity.

Timing is an important consideration when arranging for stocking and storing drywall on the job site. Job progress must be carefully monitored to be sure that the drywall sheets are placed inside the building while there is adequate access. If all exterior construction is finished and the building has been completely closed in, the access needed for stocking drywall sheets may be eliminated. In some buildings it is necessary to stock drywall sheets through access holes provided in the walls, floor, or roof. Failure to stock the job at the proper time may result in the loss of access. However, stocking drywall sheets too early can increase the potential for damage.

When drywall sheets must be stored outside the building, be sure that they are stacked flat and off the ground. The stacks should not be placed in low-lying areas where water can accumulate. The stacks must be covered on top and on all exposed sides with water-resistant sheet material. Drywall sheets readily absorb moisture and must be protected from the elements at all times. It is preferred that drywall sheets not be stored outside the building for long periods of time.

INSTALLING DRYWALL

Prior to the placement of material orders, the estimator and supervisor meet in a preplanning conference. They review the various contract documents including the subcontract, project schedule, plans and specifications, etc. Upon arriving at the site, the supervisor conducts a thorough inspection of the project and specific work areas to determine if the job is ready for drywall. Smaller projects may not warrant an on-site supervisor and the framer or drywall hanger must make the inspection. A checklist should be followed when making on-site inspection. See Figure 4-2.

If damaged or carelessly installed materials which can affect the productivity or performance of the other systems in the structure are observed, the items should be reported to their immediate supervisor or appropriate representative of the general contractor. Examples of such items include: broken or protruding plumbing pipes, loose or damaged electrical wiring, improperly installed braces, insulation batts that are missing or improperly installed, walls containing crooked studs, or walls with misaligned studs, etc.

PLANNING AND INSPECTION CHECKLIST		
Item	**Completed**	**Comments**
Inspection Card Signed		
Framing		
Rough Electrical		
Rough Plumbing		
Insulation		
Heating and Air Conditioning Ducts		
Drywall Stocked		
Nails Stocked		

Figure 4-2. A checklist should be followed when making on-site inspections.

Local building codes require that the building permit be displayed on the job site in a location where it can be easily seen. See Figure 4-3. The building permit lists all of the construction items which must be inspected and approved as the job progresses. Trades involved in applying covering materials such as insulation, drywall, lath and plaster, roofing, etc., which conceals the work of others, shall not proceed with their work until the work that is covered has been completed, inspected, and signed off by the building inspector.

POST THIS CARD ON BUILDING BEING ERECTED

WILL COUNTY
BUILDING AND USE
PERMIT
HAS BEEN SECURED

Lot 39 Sec. No. — Tax Page No. —
Subdivision LAKEVIEW Township CRETE
By OWNER Date —
Issued By Will County Building
No. 9610574

Figure 4-3. The building permit should be displayed on the job site in a location where it can be easily seen.

Only a portion of work listed on the building permit concerns the drywall worker. Framing, Rough Electrical, Rough Plumbing, Insulation, and Heating and Air Conditioning Ducts must be signed off before drywall hanging may begin. The following items should be completed before hanging drywall:

Framing

- The structure should contain all required plates, studs, joists, and backing.
- All framing members should be securely anchored.
- Braces should be properly installed and should not protrude beyond the face of the walls or ceiling.
- Studs and joists should not be badly bowed or twisted.

Rough Electrical

- Wires should not be disconnected from electrical boxes.

- Pigtails and other lengths of exposed wire used for connecting appliances should be properly located.
- Wires should not be cut or nicked.
- All boxes should be securely fastened.

Rough Plumbing

- Stub-outs for fixtures and appliances which are installed after the drywall has been completed should be the only pipes protruding beyond the wall surface.
- All stub-outs must be properly anchored.

Insulation

- All insulation batts should be in place.
- All vapor barriers should be in place.

Heating and Air Conditioning Ducts

- Ductwork for grills or registers must not protrude past the wall or ceiling face.
- All ductwork which is covered must be firmly secured.

Nail-On Installation

Most drywall for residential and light commercial installations is nailed on. This method provides the most economical means of attaching drywall sheets to wood framing members. In addition, if the drywall sheets are properly applied, the resulting wall and ceiling surfaces provide many years of trouble-free service. Factors which ensure a successful nail-on installation include:

- Employing competent workers.
- Using proper drywall nails.
- Driving nails with good dimples and proper spacing.
- Using the longest practical lengths of drywall sheets available.
- Installing drywall trim straight and true.
- Applying tape and finishing compound properly to joints and trim members.
- Removing all excess joint compound from trim and adjacent surfaces.
- Cleaning and removing all drywall debris.

Drywall Sheet Selection. When selecting drywall for a job, order the longest sheets that can be conveniently used. Preferably, the sheets should be long enough to cover the walls and ceilings from corner to corner. This eliminates the need for butt joints, which are more difficult to conceal during the taping and finishing process. In addition, because there are fewer sheets to handle and fewer joints to finish, both the drywall hanging and taping are completed in less time.

Drywall Nails. The drywall nails preferred for nail-on drywall installations have cupped heads and excellent pull-out resistance, are treated to resist rust, and penetrate 1″ into framing members. Other types of nails may hold drywall sheets in place but do not perform as well. Nails that are too long penetrate framing members more than 1″ and can cause nail pops as the wood framing members dry out and shrink. Nails with an incorrect head design cut the face paper when driven and have less holding power.

Drywall nails that are driven properly hold the drywall sheets tight against the framing members. They are spaced according to the local building code requirements and provide a good dimpling effect. See Figure 4-4. The drywall sheets should be pressed against the framing member as the nails are being driven. This helps to reduce the possibility of loose drywall, fractures, and dimples that are too deep.

Drywall nails should be driven into the center portion of the framing members. See Figure 4-5. If driven too close to the edge, they are deflected out the side of the framing member and must be removed or covered with another nail.

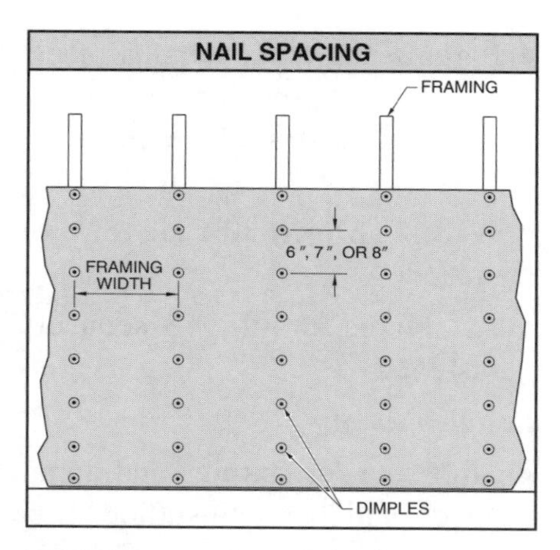

Figure 4-4. Drywall nail spacing is specified by local building codes.

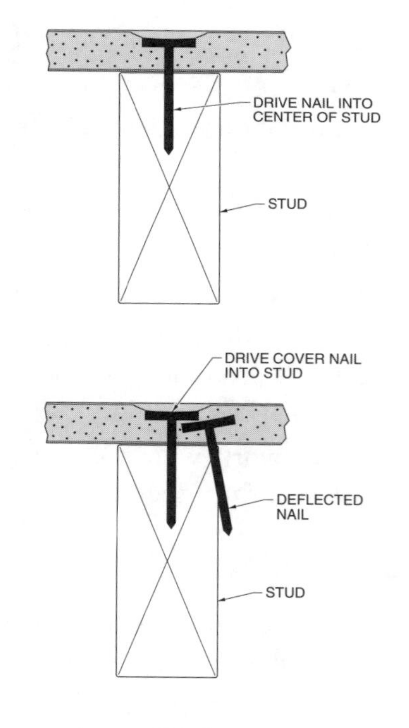

Figure 4-5. Drywall nails should be driven into the center portion of the framing members.

Nail-on drywall is generally hung from top to bottom. See Figure 4-6. Residential nail-on installations usually begin by hanging the ceiling drywall sheets first. The drywall sheet is supported by the drywall hangers until sufficient nails have been driven into the wood joists to support the weight of the sheet. When the entire ceiling has been completed, the top sheets of the walls are hung followed by the bottom sheets. The bottom sheets are lifted with a drywall lifter until they are positioned snugly (but not forced) against the bottom of the top sheets. The space remaining at the floor (approximately ½″ wide) is covered by the baseboard.

Nailing Inspection. Drywall hangers must comply with the requirements of the local building code when hanging and nailing drywall. Most building authorities require a nailing inspection before the application of the taping is permitted. Building inspectors look for the rating stamps, size of the drywall nails used, and the spacing of the nails at the ends and in the field of the drywall sheets. If the building inspector discovers any deficiencies, the drywall hanger may be required to renail the entire job.

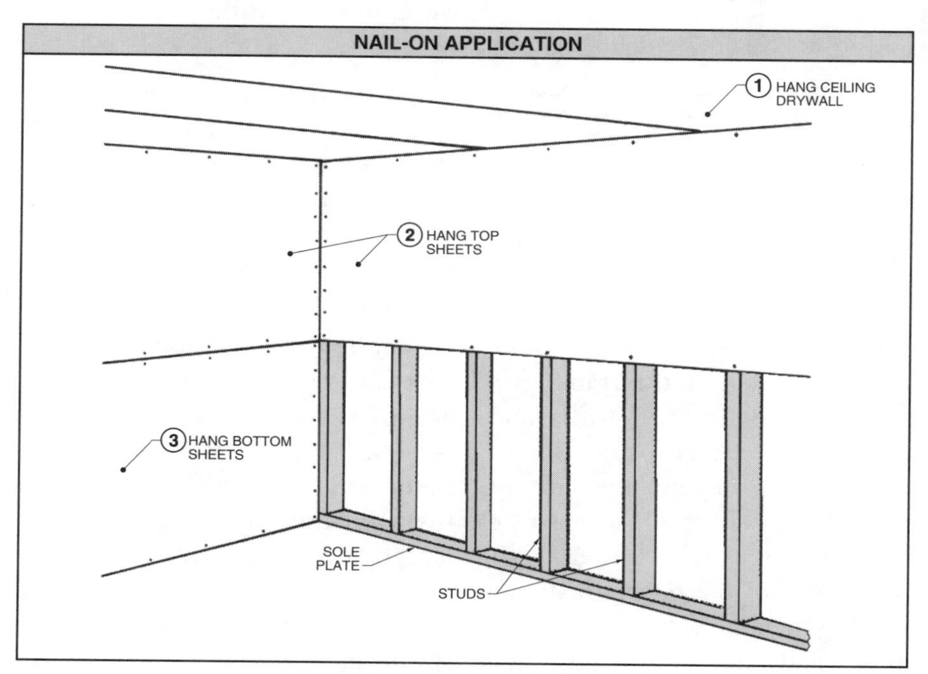

Figure 4-6. Nail-on drywall sheets are generally hung from top to bottom.

Nail-on installation is the most economical method of attaching drywall sheets to wood framing members.

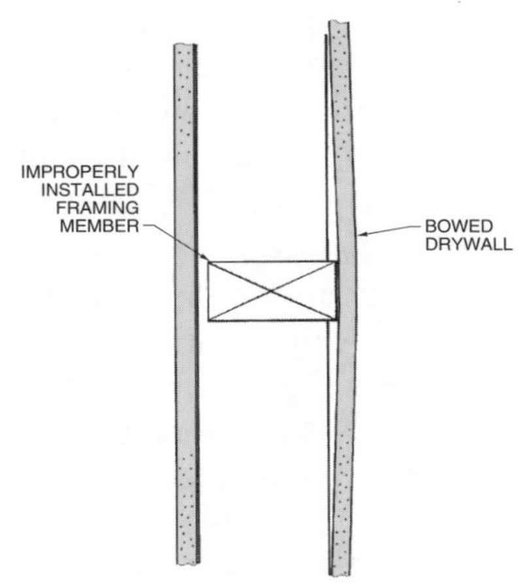

Figure 4-7. Straight framing members are essential to a quality drywall job.

Although they are not the responsibility of the drywall hanger, straight framing members are essential to a quality drywall job. See Figure 4-7. If the framing is crooked, the finished walls and ceilings will be crooked. Braces that protrude beyond the face of the wall or ceiling plane must be cut or pushed back into line. If corrections of the poorly-installed framing members are not made, humps and bulges in the finished walls and ceilings will result.

Screw-On Installation

Screw-on drywall is generally hung vertically. See Figure 4-8. Drywall sheets are screwed to metal framing members in high-rise buildings and other commercial structures where noncombustible partitions and ceiling systems are required.

A variety of metal framing members are available incorporating the use of drywall sheets attached with screws. There are similarities between nail-on and screw-on drywall installations because the same methods are used for handling, measuring, and cutting drywall sheets. However, the two methods are different in the order in which the drywall sheets are hung and the type of attachments. When attached with screws, it is common for the drywall used for the walls to be applied vertically rather than horizontally. For example, when the ceiling height is 8′-6″, 9′-0″, or 10′-0″, horizontal joints are eliminated if the drywall sheets are applied vertically.

Vertical application also produces finished wall surfaces that are free from the humps associated with end joints. Because the joint tape must be applied only at the edges of the sheets, the channels produced by the recessed edges allow the finished joint to be flush.

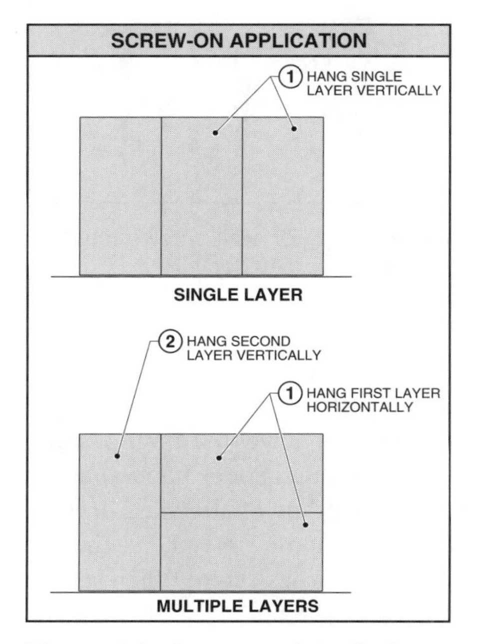

Figure 4-8. Screw-on drywall sheets are generally hung vertically.

Cut edges and ends of drywall sheets do not have recesses for the tape, creating a slight hump or bulge where the joint tape is applied. A similar condition occurs when the cut edge of a drywall sheet is placed next to the tapered edge of another sheet.

Screw-on installations may be single, double, or multiple layer. The first layer is hung horizontally and the finished layer is hung vertically. This method provides a stronger wall with increased fire resistance.

Installation Sequence. Unlike nail-on installations where the drywall sheets are applied from the ceiling down, most screw-on drywall is installed from the bottom up. This method is preferred because of the difficulty associated with holding an electric screwgun and starting the screws while holding a sheet of drywall off the floor and against the wall framing. Ceiling drywall applied with screws is usually installed after the walls have been hung. When this procedure is followed, the ceiling drywall is butted against the drywall on the walls. However, some applications require that the ceiling drywall not actually touch the walls.

The advantages of metal framing and screw-on drywall include a finished plane of walls and ceilings that is consistently straight and true. See Figure 4-9. Metal framing members do not shrink, warp, or twist as wood framing members do. When wooden studs and ceiling joists warp or twist, the nails may push through the finished drywall surface, causing nail pops. Metal framing members do not warp or twist, greatly reducing nail pops.

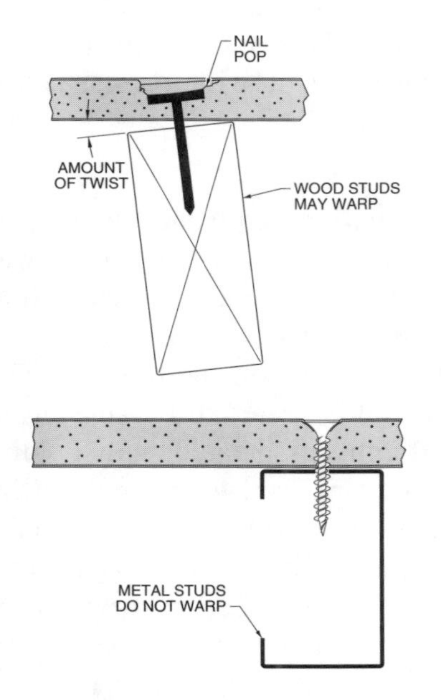

Figure 4-9. Metal studs produce a straight wall.

Screws have greater holding power than nails and require fewer fasteners than nail-on installations. Always check with the AHJ to ensure that the screws are installed with the proper spacing.

Drywall screws must be set to the proper depth. See Figure 4-10. Screws which are set too deep lose much of their holding power and do not effectively secure the drywall sheets to the framing members. Screws which are not set deep enough produce hangers. A *hanger* is a screw head that protrudes past the surface of the drywall face paper and prevents the drywall surface from being properly finished. The proper setting for drywall screws is slightly below the surface of the drywall without cutting the face paper.

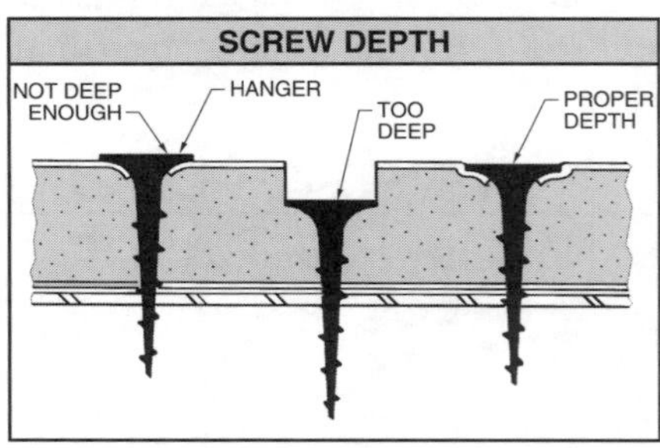

Figure 4-10. Drywall screws must be set to the proper depth.

Drywall sheets may be screwed to metal and wood framing members. Although not as popular as the nail-on method of attaching drywall to wood framing members, the use of screws is desirable. When screws are used to hang drywall sheets on wood framing members, the potential for nail pops is greatly reduced since the space caused by shrinking of wood framing members is practically eliminated. This attachment method also reduces the number of fasteners required compared to nail-on installations.

Though most architects believe drywall screws produce a higher-quality finished product, they cost approximately five times as much as nails. Although the cost of fasteners used for screw-on installation on wood framing is higher, when the cost of repairs associated with nail pops, bad joints, and crooked walls and ceilings caused by drying wood members is added, the overall costs may be similar.

Adhesive Installation

The use of adhesive for attaching drywall sheets to wood or metal framing members has gained popularity as the cost of taping and finishing has increased. Adhesive installation of drywall sheets reduces the number of fasteners (nails or screws) required and eliminates the time necessary to coat the fasteners with joint compound. Installing drywall sheets with adhesives also greatly reduces or eliminates nail pops because the adhesive eliminates the space created when wood framing members shrink. Also, fewer fasteners are required and they are installed primarily at the sheet edges, where they are covered with joint tape.

Drywall adhesive applied to wood framing members in a continuous bead improves bond strength and reduces the number of nails or screws needed to secure the drywall sheet.

Applying Panel Adhesives. Panel adhesive is supplied in tubes similar to those used for caulking compound. The manufacturer's recommendations are printed on the side of the tube and should be followed. Following these recommendations ensures a successful, trouble-free, and safe drywall installation. Costly waste can be avoided by the careful handling of the various types of adhesives. Properly installed drywall sheets, using adhesives, produce a quality wall covering with fewer fasteners. See Figure 4-11. Panel adhesive is applied in this manner:

1. The tube is inserted into a caulking gun.
2. The end of the tip is cut off the tube.
3. The inner seal of the tube is punctured.
4. The trigger is gently squeezed to start the adhesive flowing.
5. A straight bead of adhesive is applied to the center of the framing members in the center of the drywall sheet.
6. A zigzag bead of adhesive is applied to the framing members at the edges of the drywall sheet.

Always read and follow the adhesive manufacturer's directions regarding the application of the adhesive, drying time requirements, ventilation, cleanup, and disposal of adhesive residues. Use caution when handling all solvent-based adhesives. They contain flammable solvents and may be ignited by sparks or flame. The tubes used for panel adhesive are fragile and may split open if not handled with care. A split in the tube allows the adhesive to dry out, making it unusable.

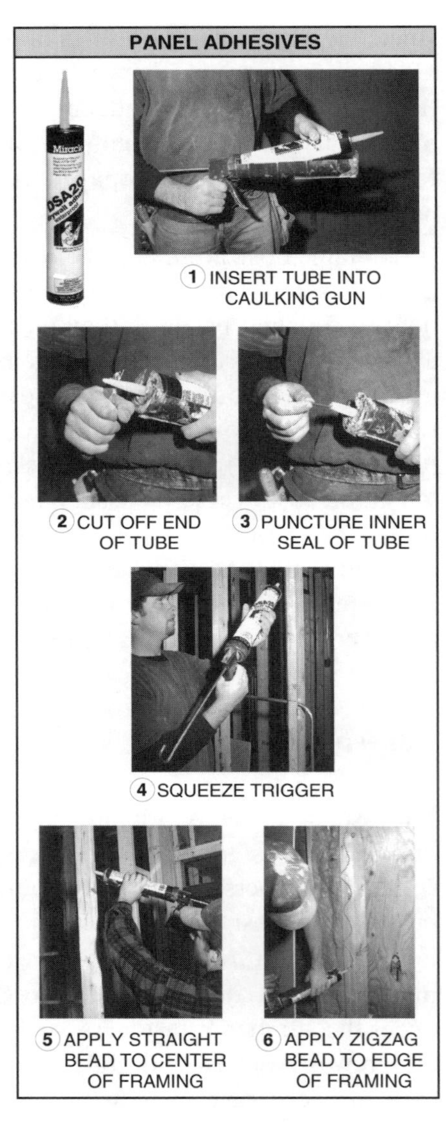

PANEL ADHESIVES
1 INSERT TUBE INTO CAULKING GUN
2 CUT OFF END OF TUBE
3 PUNCTURE INNER SEAL OF TUBE
4 SQUEEZE TRIGGER
5 APPLY STRAIGHT BEAD TO CENTER OF FRAMING
6 APPLY ZIGZAG BEAD TO EDGE OF FRAMING

Figure 4-11. Properly installed drywall, using adhesives, produces a quality wall covering with fewer fasteners.

Adhesives are commercially available for either wood or metal framing members and for both wood and metal framing members. To ensure the proper bond of drywall sheet to framing member, the correct adhesive must be selected for the application.

When using adhesive for installing drywall sheets on either wood or metal framing members, the drywall sheets should be cut to size with all cutouts completed before the adhesive is applied. When the adhesive is applied prematurely, it partially dries and glazes over. As a result, the bond between the drywall sheet and the framing member is either reduced or does not occur. After the adhesive has been applied to the framing members, the drywall sheet is positioned and pressed into place with a slight sideways motion. Screws or nails are used to attach the edges of the sheet and a few fasteners or braces are installed to ensure that adequate contact between the center of the drywall sheet and the framing members is maintained until the adhesive has thoroughly dried.

Multiple Layers of Drywall. Multiple layers of drywall may be required to achieve greater sound or fire-resistance ratings. On multiple-layer installations, the first layer can be nailed, screwed, or stapled to the wood or metal framing members. The final layer is installed with adhesive. The type of adhesive used determines the number of fasteners, if any, needed to attach the final layer in place until the adhesive dries. Another type of temporary fastener, such as double-headed nails, may be used at the center of the sheets. See Figure 4-12. These temporary fasteners are removed after the adhesive has dried. The holes are then filled with joint compound.

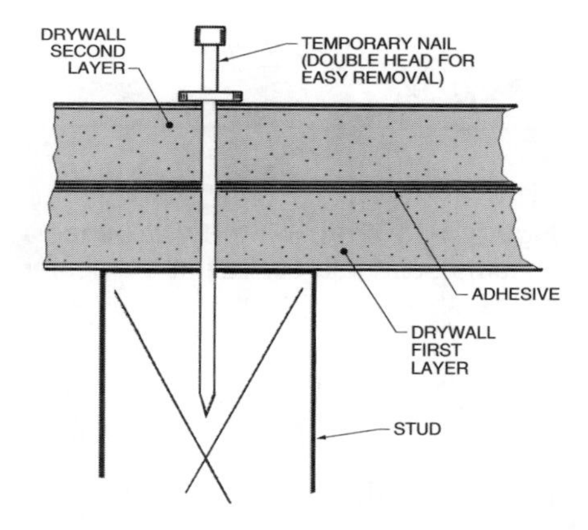

Figure 4-12. Double-headed nails may be used to temporarily hold layers until the adhesive has dried.

The need for temporary or permanent fasteners and temporary bracing in multiple-layer drywall installations when panel adhesive is used may be eliminated by prebowing the drywall sheets. Prebowing drywall sheets ensures adequate contact at the center of the sheet and a proper bond between the layers of drywall. See Figure 4-13. Prebowing is achieved by stacking the drywall sheets face up on two 4 × 4s and allowing them to remain in that position for 25 hours. This method requires some preplanning, but produces a drywall installation free from fasteners and the related taping and finishing. This installation technique is highly recommended when prefinished drywall sheets are being used.

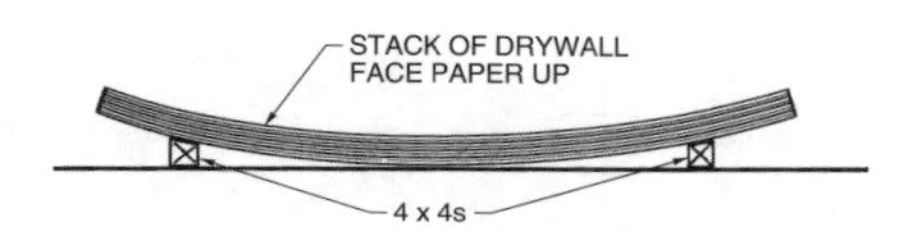

Figure 4-13. Prebowing drywall sheets ensures adequate contact at the center of the sheet.

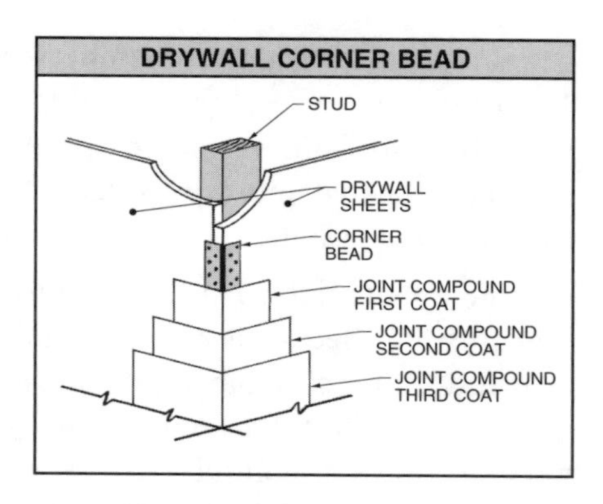

Figure 4-14. Drywall corner bead is installed on 90° outside corners which are to be finished with joint compound.

Trim Installation

Drywall corner bead, available in both square and bullnose styles, is installed on 90° outside corners which are to be finished with joint compound. See Figure 4-14. When properly installed, the two perforated flanges of the corner bead are placed tight against the two drywall surfaces forming the outside corner and aligned to form a straight line from top to bottom or end to end. Drywall corner bead may be nailed, screwed, stapled, clinched, or glued to the surface of the drywall.

Drywall corner bead finished with joint compound provides a finished square corner that is protected from damage caused by minor impact. All intersections must be properly cut to fit tightly together. When more than one piece of corner bead is required to reach from one end to the other, the two pieces must be tightly butted. Drywall corner bead must be properly aligned because it is used as a guide for the application of the joint compound. L-trim, U-trim, and J-trim are used to cap the ends or edges of drywall sheets. See Figure 4-15.

L-Trim. Drywall L-trim (L-metal) is used to cap the ends or edges of drywall sheets or as a termination when the drywall sheet meets other types of materials such as wood, steel, concrete, or masonry. L-trim may be installed using any of a variety of fasteners, depending on the application. The perforated flange of the L-trim is placed flat against the drywall face paper and is aligned straight and true before being fastened-in-place. Ends and intersections must fit tightly together. The bead portion of the L-trim must be aligned in a similar manner as corner bead.

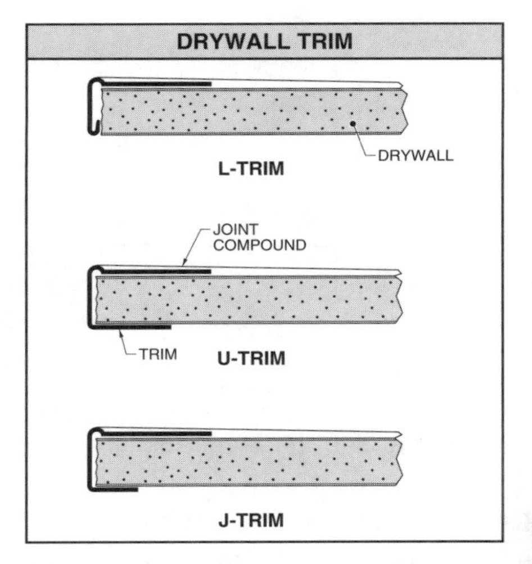

Figure 4-15. L-trim, U-trim, and J-trim are used to cap the ends or edges of drywall sheets.

U-Trim and J-Trim. U-trim and J-trim (U-metal and J-metal) are designed to encase the raw edges and ends of drywall sheets. They may be used to provide a finished termination when the drywall sheets butt against another type of material. Unlike L-trim, U-trim and J-trim do not require surface fasteners. The wrap-around design of U-trim and J-trim relies on the pressure of the drywall sheet against the framing members to hold it in place. Both of these drywall trims may be difficult to install because they must be placed on the edge or end of the sheet prior to it being installed.

U-trim and J-trim have two different types of flanges on the same member. Whether the trim is installed with the perforated flange facing out and covered with joint compound or with the smooth flange facing out and painted is determined by the details shown on the prints. Both are acceptable. However, when the perforated flange is facing out, a finished appearance similar to a finished L-trim installation is achieved.

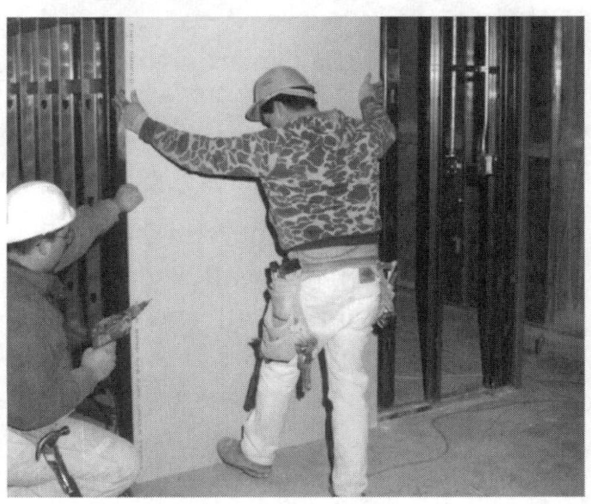

Drywall hanging begins after framing, rough electrical and plumbing, insulation, and heating and cooling ductwork are installed.

When either trim is installed on curved surfaces, both flanges must be notched at intervals from ¾″ to 6″, depending on the radius of the curve. See Figure 4-16. The notched flanges allow the metal to be bent to fit the contour of the surface. Curved applications require that the perforated flange be facing out and finished with joint compound. Both are fastened-in-place in the same manner as straight trim.

Figure 4-16. The flanges are notched when trim is applied to a curved surface.

Control-Joint Trim. Control-joint trim (expansion-joint trim) may be required when drywall sheets are installed on long wall and ceiling runs. This trim allows the drywall to expand and contract (often the result of weather conditions or building movement) without cracking of the taped joints. Expansion joints are installed at intervals indicated on the prints and in the specifications. Spaces from ½″ to ¾″ are left between the ends of the individual drywall sheets when it is installed. Control-joint trim is inserted into this space and the flanges are placed flat against the surface of the drywall and fastened to the framing members. See Figure 4-17. The flanges are coated with joint compound to produce an acceptable control joint.

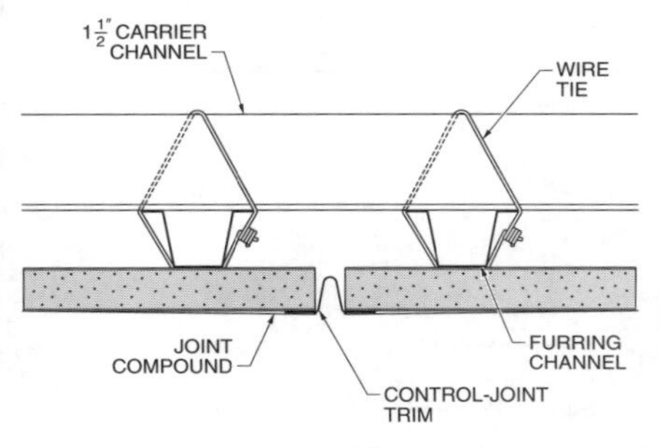

Figure 4-17. Control-joint trim allows the drywall to expand and contract without cracking the taped joints.

Reveal Trim. Reveal trim (Fry® trim) is installed when the drywall sheets are intended to have visible surface reveals as part of the finished wall or ceiling surface. See Figure 4-18. Reveal trim is installed with its perforated flanges flat against the surface of the drywall on both sides of the space provided for the reveal. It is fastened to the framing members and coated with joint compound to provide a flush surface on each side of the reveal.

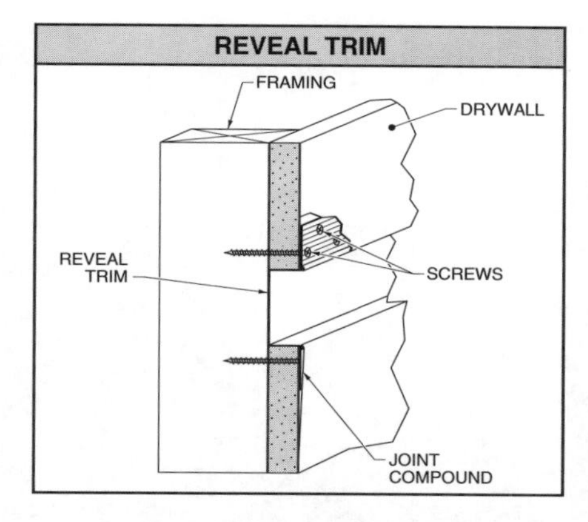

Figure 4-18. Reveal trim allows visible surface reveals as part of the finished surface.

Prefinished Panel Trim. Prefinished panel trim is considered to be a molding. See Figure 4-19. It serves a similar purpose as regular drywall trim except it is designed for use with vinyl- or fabric-covered drywall. Prefinished panel trim is installed without exposed fasteners, and it covers all exposed edges of the prefinished drywall sheets. When properly installed, these trims and moldings enhance the appearance and durability of the finished wall surface.

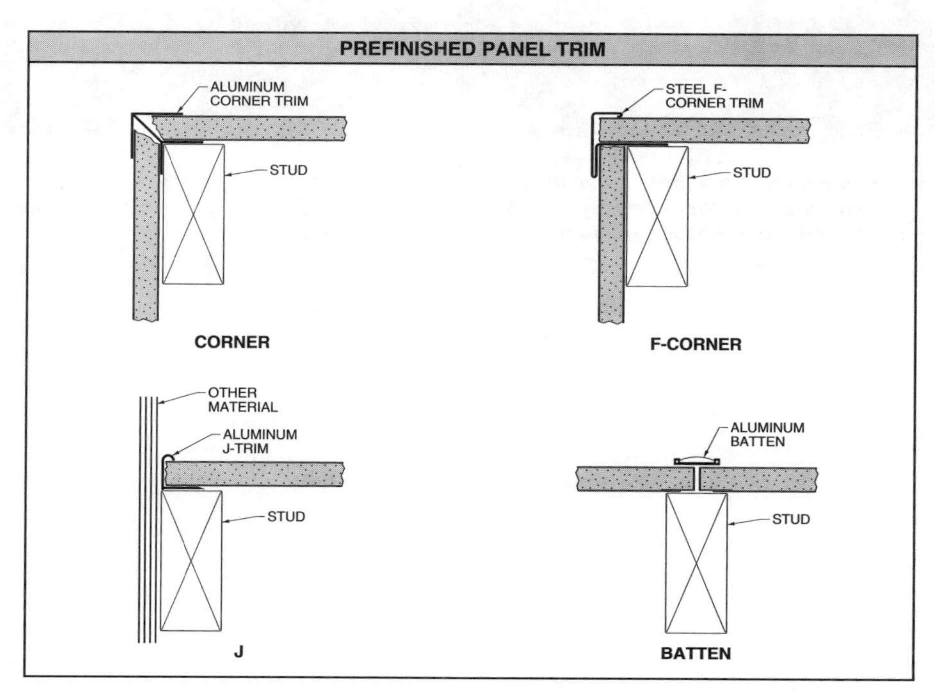

Figure 4-19. Prefinished panel trim is considered to be a molding.

Residential nail-on installations usually begin by hanging the ceiling drywall sheets first. The drywall sheet is supported by the drywall hangers until sufficient nails have been driven into the wood joists to support the weight of the sheet.

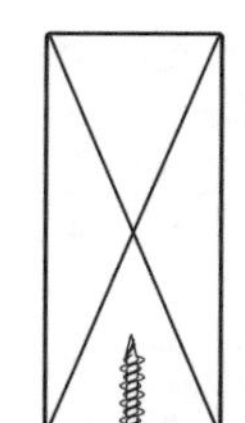

Drywall Handling and Installation

Trade Competency Test

Name ____________ Date ____________

Handling and Installation

______ **1.** Drywall nails should penetrate _____″ into the framing members.

______ **2.** Residential nail-on installations usually begin by hanging the _____ drywall sheets first.

______ **3.** Nail-on drywall is generally hung from _____.

A. top to bottom
B. bottom to top
C. left to right
D. right to left

T F **4.** Drywall nails should be driven into the center portion of the framing members.

______ **5.** Bottom sheets are hung approximately _____″ off the floor.

______ **6.** Screw-on drywall is generally hung _____.

______ **7.** In a single-layer, screw-on installation, the first layer is hung _____.

T F **8.** Nails have greater holding power than screws.

T F **9.** Drywall corner bead may be nailed, screwed, stapled, clinched, or glued to the surface of the drywall.

T F **10.** Drywall sheets should be stocked by leaning them against a wall.

______ **11.** A(n) _____ is a screw head that protrudes past the surface of the drywall face paper and prevents the drywall surface from being properly finished.

_______________ **12.** Drywall corner _________ is installed on 90° outside corners.

_______________ **13.** Prefinished panel trim is considered to be a(n) _________.

T F **14.** L-trim requires surface fasteners.

T F **15.** U-trim and J-trim require surface fasteners.

Prefinished Panel Trim

_______________ **1.** Batten

_______________ **2.** F-corner trim

_______________ **3.** J-trim

_______________ **4.** Corner trim

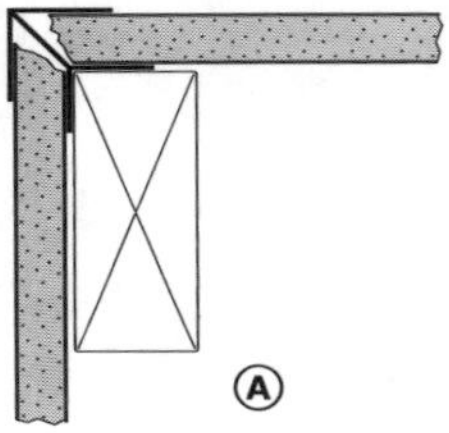

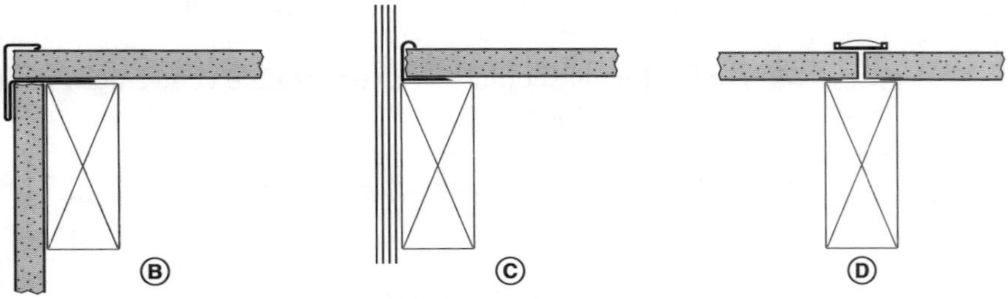

Ⓐ Ⓑ Ⓒ Ⓓ

Trim

_______________ **1.** L-trim

_______________ **2.** U-trim

_______________ **3.** J-trim

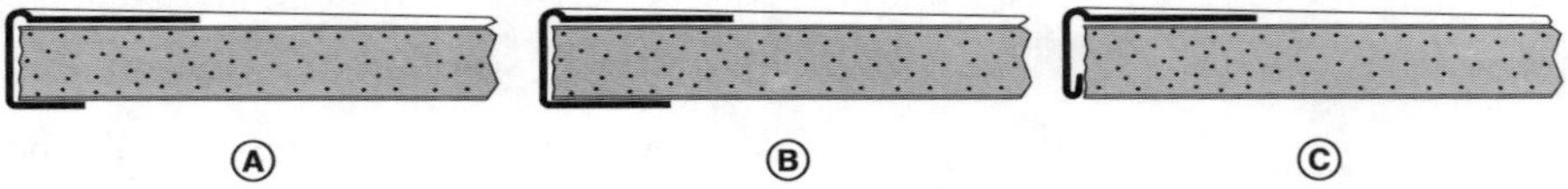

Ⓐ Ⓑ Ⓒ

5 Metal Framing Assemblies

Light-gauge metal studs are used to frame partitions and heavy-gauge metal studs are used to frame load-bearing and nonload-bearing exterior walls. Suspended drywall ceilings are constructed with drywall furring channels. Shaftwalls use drywall to provide required fire ratings.

METAL STUD WALLS

The walls and partitions of commercial buildings are commonly constructed with metal studs. Light-gauge metal studs are used for partition framing. Heavy-gauge metal studs are used for load-bearing and nonload-bearing exterior walls.

Metal studs are economical, noncombustible, and lightweight. Light-gauge metal partitions can be quickly installed and removed. Heavy-gauge metal stud walls provide a system that is structurally sound.

Light-Gauge Metal Stud Walls

Noncombustible light-gauge metal studs are used for wall or partition framing in malls, strip centers, and high-rise structures. Prior to beginning construction of light-gauge metal stud partitions, the framer should examine the material supplied to determine whether or not it complies with the construction plans and specifications. The width, length, gauge thickness, manufacturer, and quantity of metal framing material should be checked to confirm that a sufficient number of properly-sized studs and tracks have been supplied to complete the work.

Layout and Track Installation. Six layout and track installation steps are recommended and should be followed when constructing light-gauge metal stud partitions:

1. Determine the location of the partitions from the floor plan or from the reference marks provided.
2. Mark the locations on the floor or confirm that the layout provided is accurate. Use a chalk line to mark the locations of the partition tracks on the floor.
3. Transfer the layout to the ceiling grid or overhead structure using a plumb bob or laser level.
4. Lay out the locations of all doors, windows, and other openings. Clearly mark these locations on the floor.
5. Fasten the tracks to the ceiling or overhead structure.
6. Fasten the tracks to the floor.

Malls, strip centers, and high-rise buildings use light-gauge metal studs for wall or partition framing.

Tracks are usually fastened to the ceiling or overhead structure before they are fastened to the floor to allow easy access for the rolling scaffold and to prevent damage to the bottom tracks. Fasteners used to attach the tracks must be the specified type, length, and diameter and must be placed at the intervals indicated on the plans, listed in the specifications, or as determined by the AHJ.

Stud Installation. Seven installation steps are recommended and should be followed when constructing light-gauge metal stud partitions:

1. Lay out the stud spacing on the top and bottom tracks. Studs are placed at the intervals indicated on the plans, listed in the specifications, or as determined by the AHJ.
2. Measure the floor-to-ceiling distance between the top and bottom track to determine the stud length.
3. Cut studs approximately ½″ shorter than the floor-to-ceiling dimension. See Figure 5-1.
4. Stand the studs within the top and bottom tracks and twist them into place. The stud spacing must be accurately maintained to avoid unnecessary cutting of the drywall sheets.

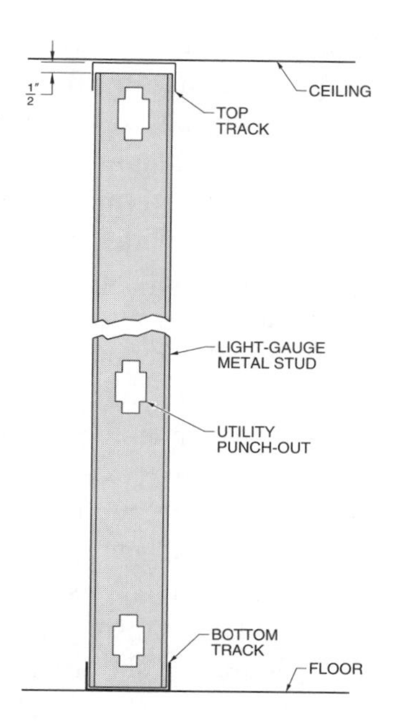

Figure 5-1. The studs are cut approximately ½″ shorter than the floor-to-ceiling dimension.

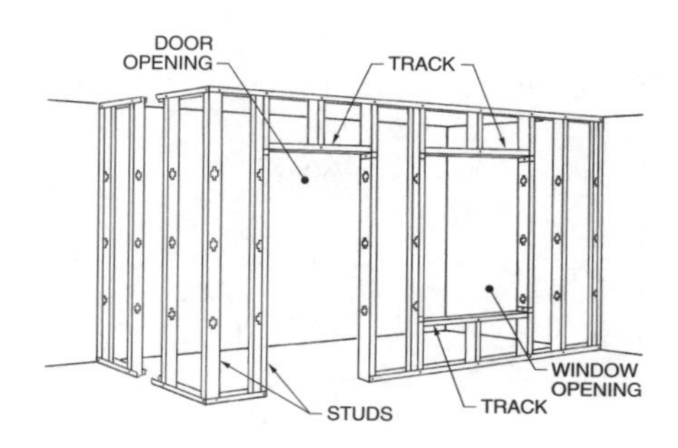

Figure 5-2. Door and window openings are framed with stud tracks.

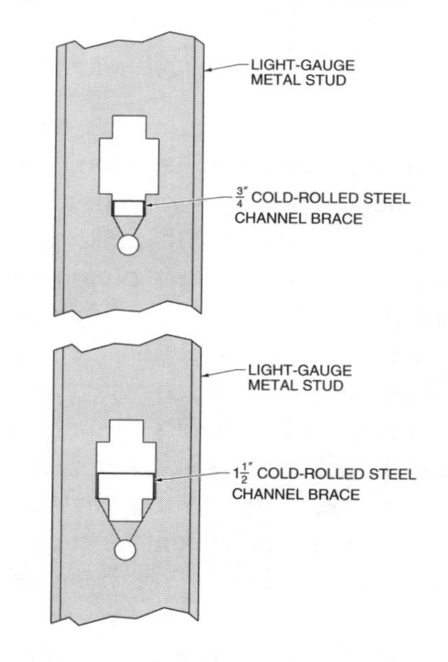

Figure 5-3. Door and window studs may require ¾″ or 1½″ cold-rolled steel channel braces.

5. Frame door, window, and other openings with tracks at the top of door openings and at the top and bottom of window and other openings. See Figure 5-2. Heavy-gauge studs may be required for the sides of door and window openings. This must be determined from the plans or specifications before proceeding. Door and window studs may also require ¾″ or 1½″ cold-rolled steel channel braces. See Figure 5-3.

6. Anchor each stud to the top and bottom tracks with self-drilling, self-tapping framing screws. See Figure 5-4.

7. Install partition bracing or backing as required.

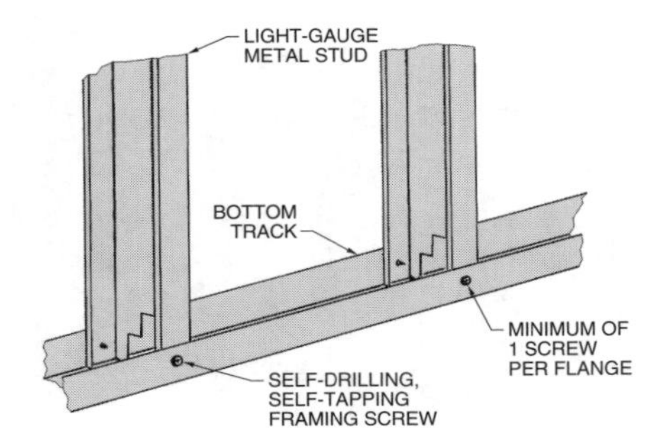

Figure 5-4. Light-gauge metal studs are attached to the top and bottom tracks with self-drilling, self-tapping framing screws.

Framing Inspections. Upon the completion of the partition framing, check all partitions to make sure that they are straight and plumb and that the stud spacing is accurate. When framing inspections are required, they will be requested after the other tradesworkers (plumbers, electricians, etc.) have completed their work. This ensures that any cuts made in the framing members by these workers have been repaired and that the framing is intact. The drywall sheets are applied to the partition framing after the other tradesworkers have completed their work and the work has been inspected.

Heavy-Gauge Metal Stud Walls

Noncombustible heavy-gauge metal studs are used extensively for both load-bearing and nonload-bearing exterior wall framing. Prior to beginning the construction of any heavy-gauge metal stud walls, the framers should examine the material supplied to determine whether or not it complies with the construction plans and specifications. Heavy-gauge framing components shall comply with the requirements because the structural integrity of the building is dependent on properly framed and braced exterior walls. The name of the manufacturer, width, length, and gauge thickness of all studs and tracks must be checked. The quantity of metal framing material should also be checked to confirm that sufficient studs and tracks have been supplied to complete the work.

Layout and Track Installation. Six layout and track installation steps are recommended and should be followed when constructing heavy-gauge metal stud walls:

1. Determine the location of the walls from the floor plan or reference marks provided.
2. Mark the locations on the floor or confirm that the layout provided is accurate. Use a chalk line to mark the location of all wall tracks on the floor.
3. Transfer the layout to the overhead structure using a plumb bob or laser level.

4. Lay out the locations of all doors, windows, and other openings. Clearly mark these locations on the floor.
5. Fasten the tracks to the overhead structure.
6. Fasten the tracks to the floor.

The overhead tracks are usually fastened to the structure before the bottom tracks are fastened to the floor for two reasons:

1. The spray-applied fireproofing should be installed after the top tracks are in place to ensure an adequate seal between the walls and the decking or beams.
2. Installing the top tracks first allows access for rolling scaffolds and reduces the damage to the bottom tracks.

Fasteners used to attach the tracks must be the specified type, length, and diameter and must be placed at the intervals indicated on the plans, listed in the specifications, or as determined by the AHJ. If welds are specified for attaching the top tracks to the overhead structure, they must be the proper length, width, spacing, and locations indicated on the plans or listed in the specifications.

Stud Installation. Six installation steps are recommended and should be followed when constructing heavy-gauge metal stud walls:

1. Lay out the stud spacing on the top and bottom tracks.
2. Measure the distance between the top and bottom track to determine the stud length. If the wall is load-bearing, the studs must be full-length and must be properly seated in the top and bottom tracks. See Figure 5-5. When a nonload-bearing wall is being constructed, studs may be cut approximately ½″ shorter than the floor-to-ceiling dimension.
3. Place the studs within the top and bottom tracks and twist into place. Stud spacing must be accurately maintained to ensure that the design criteria are met and to avoid unnecessary cutting of the drywall.

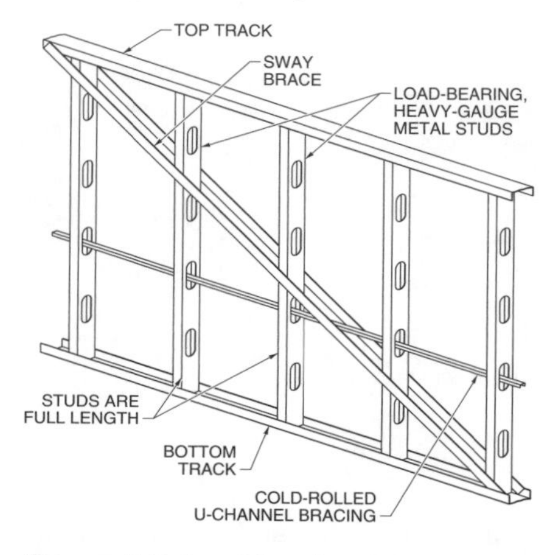

Figure 5-5. Load-bearing, heavy-gauge metal studs must be full-length.

Drywall is installed on metal stud wall framing after other tradesworkers have completed their work and framing inspections have been performed.

4. Frame door, window, and other openings with tracks or box headers at the top of the openings and track at the sills of windows and other openings. See Figure 5-6.

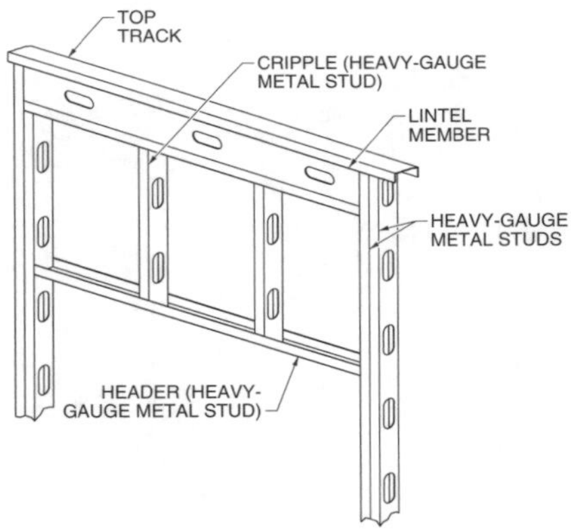

Figure 5-6. Door and window openings are framed with heavy-gauge metal studs.

Heavier gauge studs may be required for the sides and headers of door and for window openings. Check the plans and/or specifications to determine stud sizes. Door and window studs may also require ¾″ or 1½″ cold-rolled steel channel braces. See Figure 5-7.

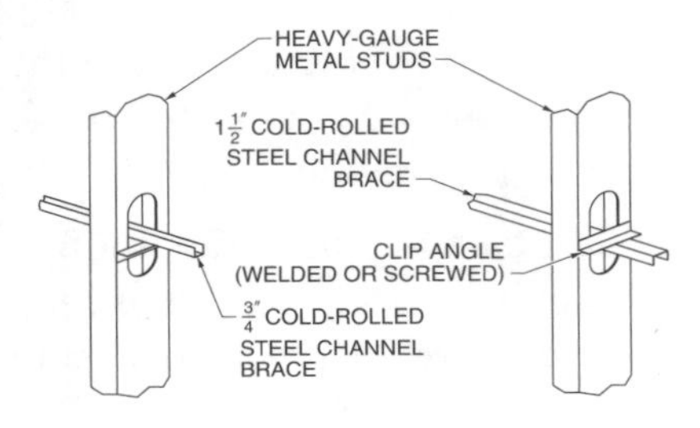

Figure 5-7. Door and window studs may require ¾″ or 1½″ cold-rolled steel channel braces.

5. Securely anchor studs to the top and bottom tracks with self-drilling, self-tapping framing screws or welds. See Figure 5-8.
6. Be sure all wall bracing and backing is installed as required.

Framing Inspections. After completing the wall framing, check to be sure that all walls are straight and plumb and the stud spacing is accurate. Framing inspections are required for heavy-gauge walls. They are called for after the other tradesworkers have completed their work. This ensures that all cuts made to the framing members have been repaired and the framing is structurally intact. The drywall and gypsum sheathing is installed on the wall framing after the other tradesworkers have completed their work and the walls have passed inspection.

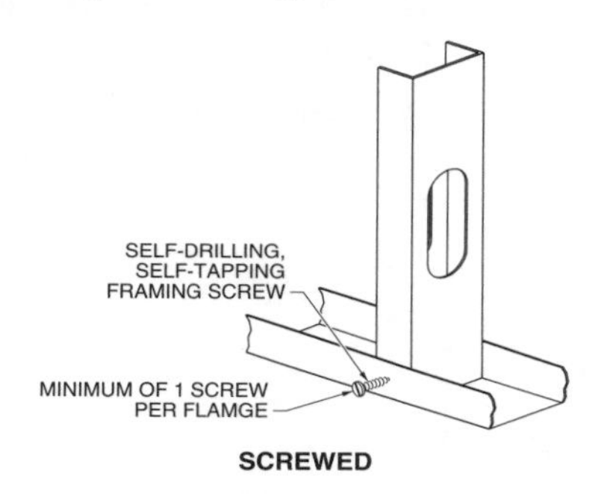

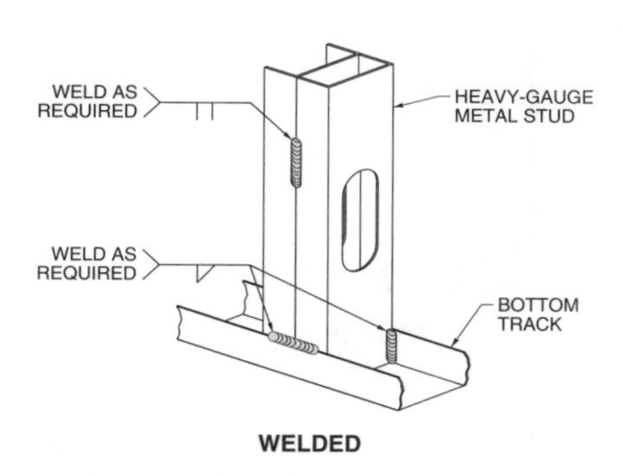

Figure 5-8. Heavy-gauge metal studs are attached to the top and bottom tracks with self-drilling, self-tapping framing screws or welds.

DRYWALL FURRING CHANNEL CEILINGS

Suspended drywall ceilings constructed with drywall furring channels are used in high-rise buildings and in other types of commercial structures. The light-gauge metal framing components provide a lightweight, noncombustible framing system suitable for drywall application. The ceiling framing assembly consists of drywall furring channels which are wire tied, screwed, or clipped to 1½″ cold-rolled steel carrier channels. The carrier channels are suspended from the overhead structure of a new building or from the existing ceiling framing of an old building using No. 8 hanger wires. See Figure 5-9.

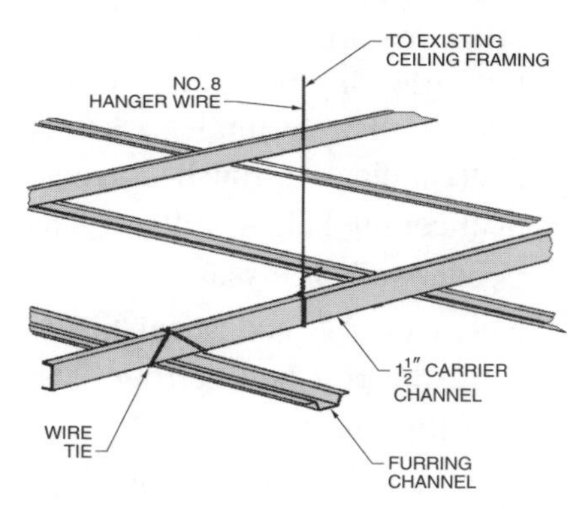

Figure 5-9. The carrier channels are suspended from the overhead structure of a new building or from the existing ceiling framing of an old building using No. 8 hanger wires.

The following six steps are recommended and should be followed when constructing suspended furring channel ceilings, providing the hanger wires have been previously installed:

1. Determine the finished ceiling height from the plans. The *finished ceiling height* is the height of the ceiling after the drywall sheets and any other covering have been applied.

2. Subtract the thickness of the drywall sheets and any other covering to determine the rough ceiling height. The *rough ceiling height* is the height of the ceiling before the drywall sheets and any other covering have been applied. This is necessary for determining the finished ceiling height, since other materials such as acoustical tile may be applied to the drywall. For example, a set of plans indicate a finished acoustical ceiling height of 9′-6″. The ⅝″ drywall sheets are to be covered with ¾″ acoustical tile to form the finished ceiling. The ⅝″ thickness of the drywall sheets added to the ¾″ thickness of the acoustical tile equals a total thickness of 1⅜″. When 1⅜″ is added to the 9′-6″ finished ceiling height, the rough ceiling height is 9′-7⅜″.

3. Measure and mark this location around the room with the aid of a laser or water level. This is the height to the bottom of the furring channels.

4. Raise the laser ⅞″ to allow for the thickness of the furring channels and place a 90° bend on each hanger wire.

5. Lay the 1½″ cold-rolled steel carrier channels in the bend of the hanger wires. See Figure 5-10. Hanger wires are spaced at intervals determined from the plans or as required by the AHJ.

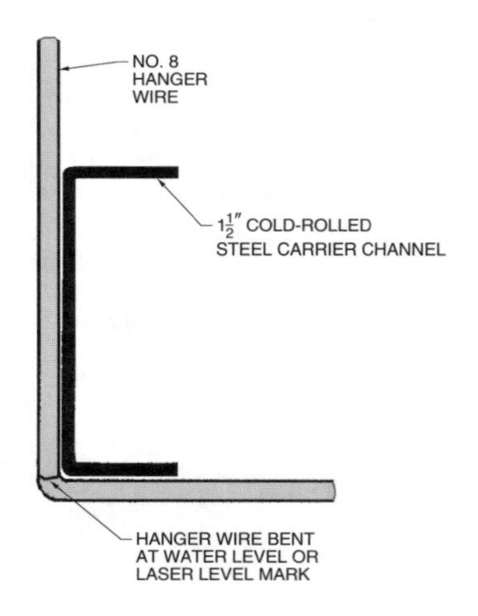

Figure 5-10. The carrier channels are laid in the bend of the hanger wires.

6. Wrap the channel twice and twist the wire around itself. See Figure 5-11. Repeat this step until all carrier channels have been installed and all hanger wires have been wrapped and twisted.

7. Lay out the spacing of the furring channels on the carrier channels.

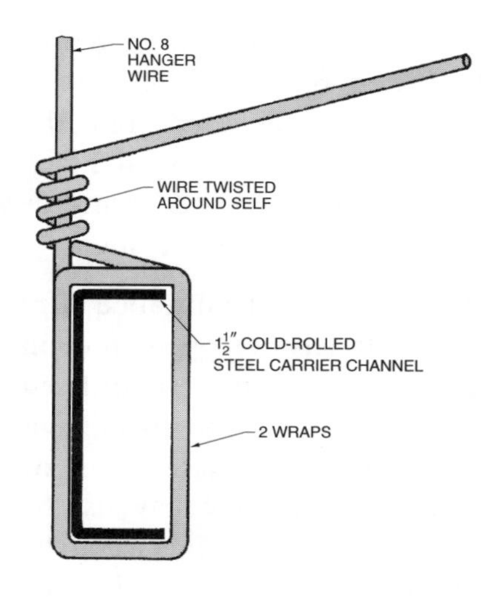

Figure 5-11. The hanger wire is wrapped around the channel twice and then twisted around itself.

8. Install the furring channels by attaching them to the carrier channels using wire ties, clips, or framing screws. See Figure 5-12. The furring channels are installed at intervals indicated on the plans or according to the AHJ. The furring channels must be accurately positioned to reduce unnecessary cutting of the drywall sheets.

When the framing is completed, the plane of the ceiling should be checked by sighting along the bottoms of the furring channels. Adjustments may be made to raise or lower areas that are out of alignment by striking the carrier channel with a hammer to lower the framing or by sliding the hanger wire to one side to raise the channel. See Figure 5-13.

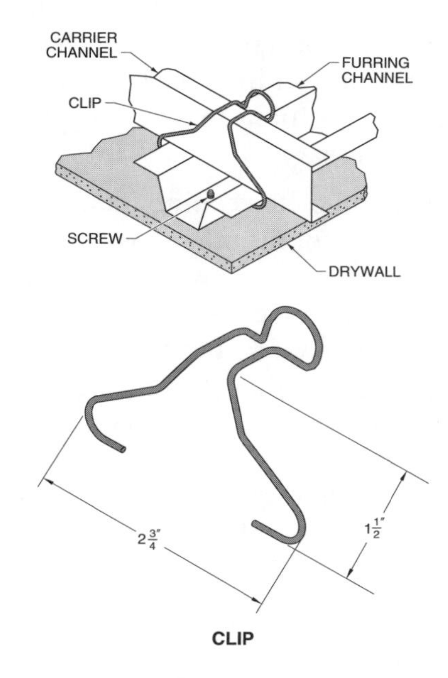

Figure 5-12. Clips are used to attach the furring channel to the carrier channel.

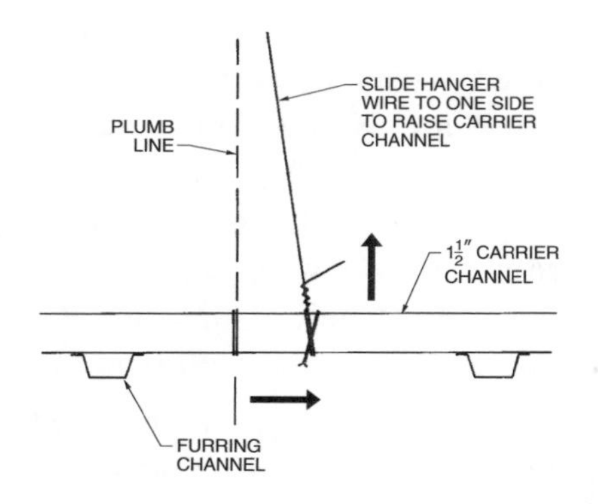

Figure 5-13. The carrier channel may be raised by sliding the hanger wire to one side.

After the alignment adjustments are completed, any hanger wires that have been moved away from dead center are secured with a wire tie or framing screws. This prevents them from sliding back to their original position and causing the ceiling to sag.

The completed ceiling framing is now ready for the other tradesworkers to perform their work. After the other tradesworkers have completed their work, the framer must inspect the ceiling to be sure that the members are intact and properly aligned, and the layout spacing has been maintained. If inspections are required by the AHJ, they may be ordered at this time.

Hanger Wires

The use of wires for supporting suspended ceilings is the most common practice employed for both new and existing buildings. Hanger wires used for ceiling framing are No. 8 gauge galvanized steel wire. They are available in lengths from 4′ to 20′ and in 100 lb rolls. Hanger wires may be installed either before or after the concrete floors have been poured.

Wires intended to be installed before the concrete is poured are provided with a pigtail on one end. The insertion of a short length of reinforcing steel in the pigtail of the hanger wire may be required to provide added pull-out resistance. Prior to the pouring of concrete, the wires are dropped through holes drilled in the plywood forms or punched in the metal decking.

Hanger wires to be installed after the concrete has been poured or in existing buildings may be ordered with clips suitable for attachment with a powder-actuated gun or drilled fasteners. They may also be ordered in straight lengths when the installation requires that they be connected to bar joists or other overhead structural members.

Whether the hanger wires are installed before or after the concrete is poured, the location and spacing is determined prior to placement. The plans may indicate the direction of the drywall furring channels, or the presence of recessed light fixtures may dictate the direction. After determining the direction of the furring channels from the plans or the layout of the recessed ceiling lights, the location of the carrier channels, which are installed perpendicular to the furring channels, can be determined. The spacing of the carrier channels and hanger wires is provided on the plans or in the specifications. This spacing must be maintained to ensure the wires are installed at the intervals indicated in both directions.

When hanger wires are installed through plywood forms prior to the concrete pour, they must be straightened after the forms have been removed. The condition of the wires after the forms have been removed is the result of the workers removing the forms and bending the wires out of their way. The carrier channels used for suspended ceiling framing must not be installed with bent hanger wires. Doing so produces a ceiling that becomes completely out of alignment as the wires eventually straighten from the weight of the ceiling assembly.

Straightening hanger wires is accomplished by applying a downward pressure on each wire while twisting it three or four turns. A tool for this purpose may be fashioned from a mechanic's speed handle. A hole with sufficient diameter to accommodate the hanger wire is drilled in the square end. Approximately 1½″ of the lower end of the wire is bent 90° and inserted in the hole previously drilled in the speed handle. The tool is pulled downward with one hand as it is being rotated with the other hand. This pulling, twisting motion straightens the wires. Do not overtwist the wires. Doing so causes the wire to break off at the surface of the concrete.

Ruud Lighting, Inc.

Soffits are used to enclose the area above cabinets or other fixtures and may be constructed from metal studs.

SOFFITS AND LIGHT TROUGHS

A *soffit* is an area of the ceiling that is constructed at a finished height lower than the main ceiling. See Figure 5-14. Soffits are commonly located at the perimeter of the room. They often enclose the space above cabinets or other fixtures which do not extend to the full height of the ceiling.

Soffits may be constructed inside, outside, or inside and outside of the building. See Figure 5-15. Some soffits are designed to provide the impression the exterior surface continues into the inside with only a glass wall separation.

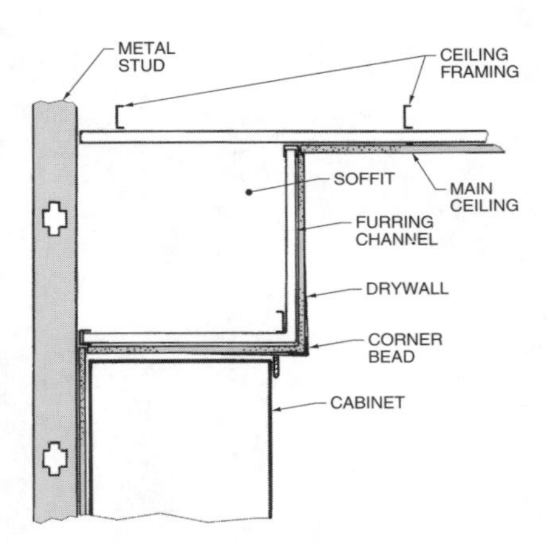

Figure 5-14. A soffit is an area of the ceiling that is constructed at a finished height lower than the main ceiling.

Soffits may be constructed outside a building.

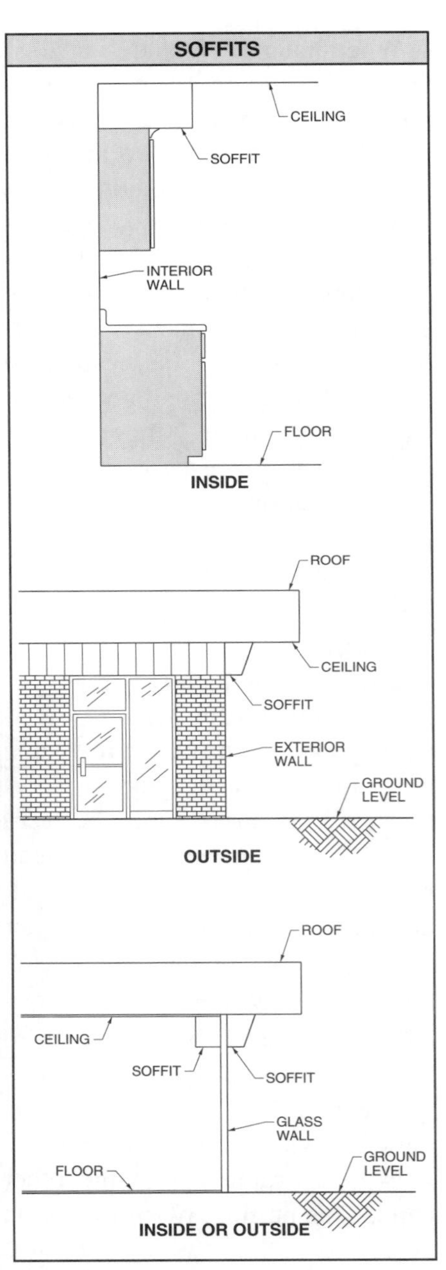

Figure 5-15. Soffits may be constructed inside, outside, or inside and outside of the building.

A *light trough* is a recessed area in the ceiling which provides a concealed location for the installation of light fixtures used for indirect lighting. See Figure 5-16. Light troughs are frequently located at the perimeter of a room and may project beyond the wall plane. Some light troughs are constructed with a wide face designed to conceal the lights installed on top of the light trough.

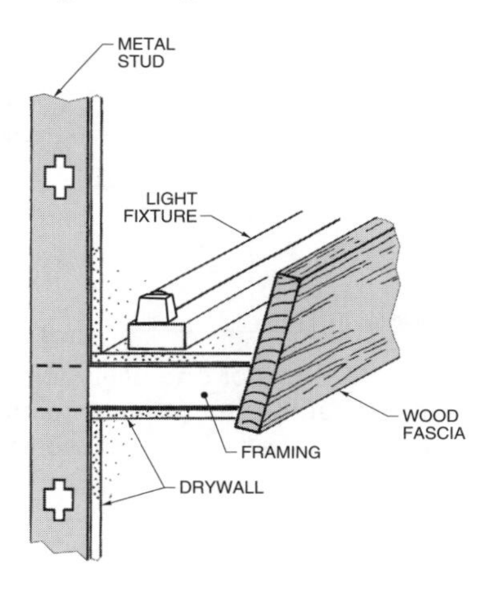

Figure 5-16. A light trough is a recessed area in the ceiling which provides a concealed location for the installation of light fixtures used for indirect lighting.

Both soffits and light troughs are usually constructed after the main wall and ceiling framing have been completed. The construction details for framing soffits and light troughs are provided on the plans. Frequently, soffits and light troughs are designed with unusual shapes and framing combinations.

Special metal shapes may be required when framing soffits and light troughs. These may be ordered from a specialty metal fabricator. Cold-rolled steel channels may be bent to form curves or cut and welded to form angles. See Figure 5-17. Combinations of stud sizes and thicknesses may be used in the construction of the soffit or light trough. The various components may be supported by hanger wires and $1\frac{1}{2}''$ channels. The bending, cutting, or welding of the components used in the fabrication may require the use of jigs or other patterns. It may be necessary for the various soffit components to be laid out on the floor. When this procedure is followed, the various members are properly shaped and then raised into place, assembled, and secured to the framing.

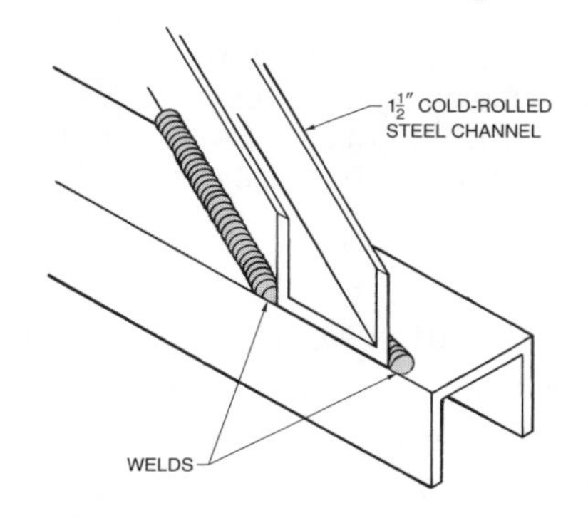

Figure 5-17. Cold-rolled steel channels may be bent to form curves or cut and welded to form angles.

SHAFTWALLS

A *shaftwall* is a rated enclosure which encloses elevators, air ducts, plumbing pipes, electrical wires, or other items which pass through the floors of a high-rise building. See Figure 5-18. All elevator and utility shafts must be constructed so that a fire originating within the shaft does not spread to the floors above. See Figure 5-19.

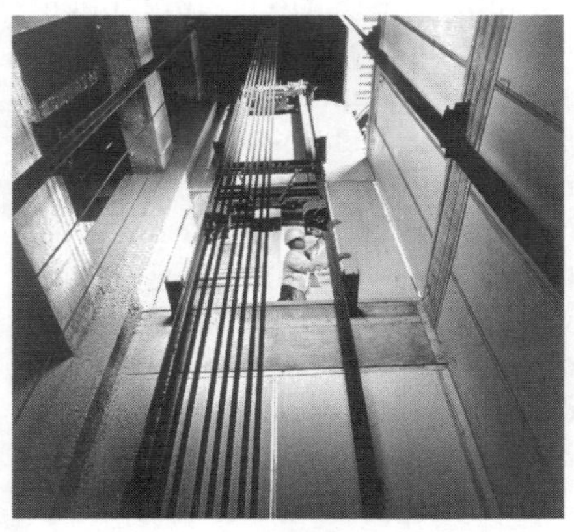

United States Gypsum Company

Figure 5-18. A shaftwall is a rated enclosure which encloses elevators, air ducts, plumbing pipes, electrical wires, or other items which pass through the floors of a high-rise building.

Shaftwall systems are designed to provide one-, two-, and three-hour fire protection. An enclosure rated for one hour is designed to allow a fire to burn for one hour on one side of the enclosure without breaking through to the other side. Manufacturers of shaftwall systems have specific designs which have been tested and approved to obtain the required rating.

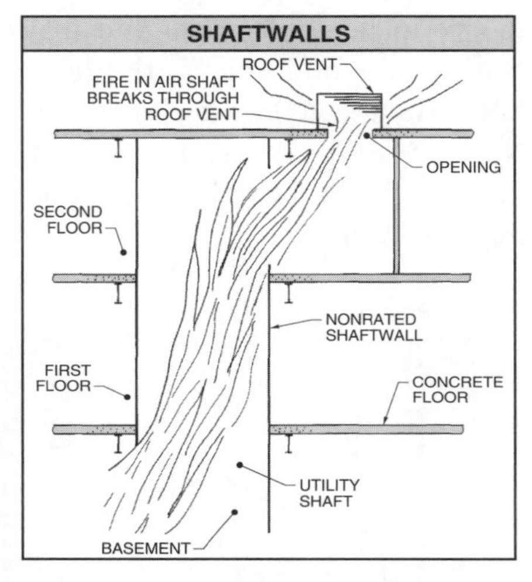

Figure 5-19. Elevator and utility shafts are constructed so that a fire originating within the shaft does not spread to the floors above.

All shaftwall systems are constructed using similar components. They incorporate the use of 1″ thick drywall sheets (coreboard) and one or more layers of drywall, depending on the fire-resistance rating required. Shaftwall systems use a variety of metal framing components. See Figure 5-20. These include metal tracks at the top and bottom, metal studs of various shapes, and metal shapes for starting or terminating the shaftwall.

National Gypsum Company

Figure 5-20. Shaftwall systems use a variety of metal framing components.

Shaftwall Installation

The following four steps are recommended and should be followed when constructing shaftwalls:

1. Follow the manufacturer's instructions. Do not substitute components of one manufacturer for components of another. Fire-resistive rating tests are conducted on assemblies constructed with the combined components from a single source. The performance of the assembly can be ensured only if the correct shaftwall components have been used.

2. Do not leave gaps between the coreboard and the walls, beams, or columns they butt against. See Figure 5-21. Although gaps will be concealed by the drywall, they form weak spots in the system and may allow fire to penetrate to the column or beam at that location.

3. When applying two or more layers of drywall, be sure that all layers are attached to the framing members with the proper fasteners and at the specified intervals. Multiple layers of drywall in shaftwalls must be offset at least 6″. See Figure 5-22.

Do not spot-screw the base layers of drywall. See Figure 5-23. Spot-screwed base layers in multilayer, fire-rated assemblies may result in the failure of the assembly due to the lack of fasteners, should the outer layer be destroyed in a fire.

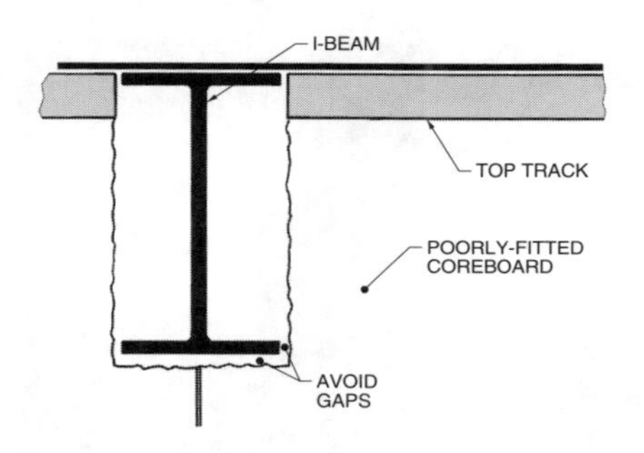

Figure 5-21. Gaps should not be left between the coreboard and the walls, beams, or columns they butt against.

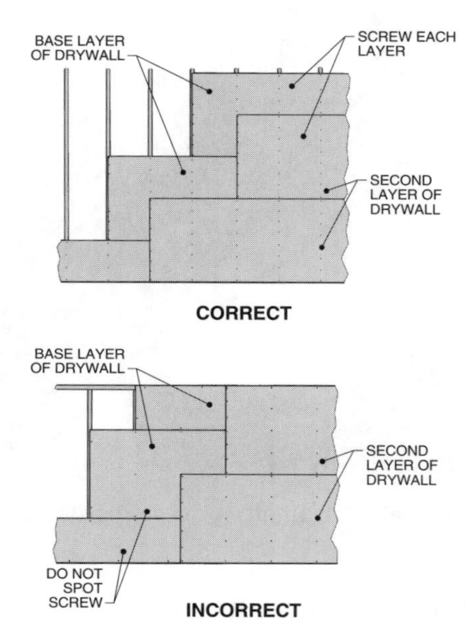

Figure 5-23. The base layers of drywall should not be spot-screwed.

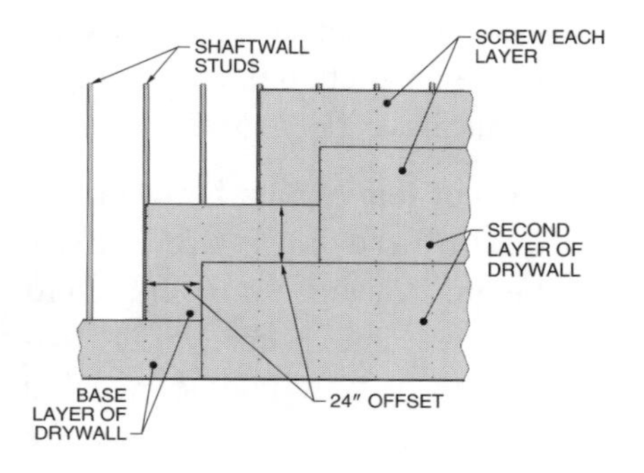

Figure 5-22. Multiple layers of drywall in shaftwalls must be offset at least 6″.

4. Follow the plans and specifications or the local building code requirements for offsetting sheet edges and ends between layers. Although the final layer of drywall conceals the underlayers, the entire assembly may fail to provide the needed protection if it is not properly constructed.

DEMOUNTABLE PARTITIONS

A *demountable partition* is a partition designed to be assembled, disassembled, and reassembled in another location with minimal damage to the partition components. Partitions of this type are most commonly used in office buildings where space requirements frequently change. Unlike standard metal stud and drywall partitions, which must be disposed of when removed, demountable partitions can be easily relocated with minimal waste.

The salvage rating signifies the percentage of partition components that are reusable when the demountable partition system is relocated. A demountable partition's salvage rating may be the determining factor when a building owner selects the type to be used. A partition with a high salvage rating may be more expensive initially; however, when the cost of the components destroyed when it is moved from one location to another is factored in, it may cost less. For example, a demountable partition with a high salvage rating that is moved from the third floor to the second floor within the same building will require a minimum number of new components. This reduces the high cost of material delivery and stocking when access is poor.

All demountable partitions, regardless of the manufacturer, are designed to divide space. Most utilize tracks or runners at the top and bottom, studs, and drywall sheets or prefinished panels that conceal the framing. See Figure 5-24. Some demountable partitions use clips that hang the panels on the studs. Others require screws for attaching the panels to the studs and rely on battens to cover the screws at the joints. There are progressive systems in which interlocking sandwich panels are started at one side of the room. Additional panels are added as the installation progresses to the opposite end.

Demountable Partition Installation

Regardless of the manufacturer or the type of system used, seven basic rules should be followed when installing demountable partitions:

1. Be sure that the partition layout has been completed before the panels are stocked in the room where they will be installed. Demountable partitions are usually installed in finished rooms and material storage space may be limited. Repeated moving of prefinished panels increases the potential for damage.

2. Double-check the layout before fastening the tracks to the floor. Damaged carpets or floor tile caused by improperly located floor tracks may be impossible to repair.

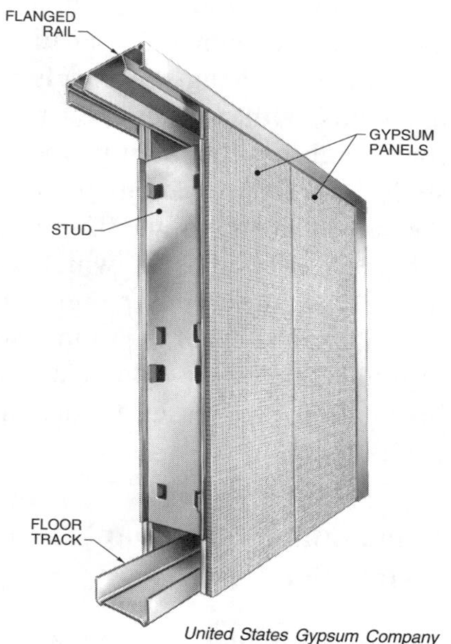

United States Gypsum Company

Figure 5-24. A demountable partition is a partition designed to be assembled, disassembled, and reassembled in another location with minimal damage to the partition components.

3. Take all measurements and make all cuts accurately. Most demountable partitions use a top track component that is a finished member with no additional trim at the ceiling. Poorly-fitting corners and track joints are unacceptable and costly to correct.
4. Install all framing members correctly. Accurate layout of the studs is very important since the prefinished panels should not be cut.
5. Handle the panels carefully to avoid damaging the panel faces and edges. The finished appearance of the partition is directly affected by the condition of the panels.
6. Be sure all door and window frames are installed correctly (plumb, true, and securely fastened to the framing).
7. Exercise care when removing unused material or scrap from finished rooms.

Metal Framing Assemblies

Trade Competency Test

Name ______________________ Date ______________

Metal Framing

______________ **1.** The _________ ceiling height is the height of the ceiling after the drywall sheets and any other covering have been applied.

______________ **2.** The _________ ceiling height is the height of the ceiling before the drywall sheets and any other covering have been applied.

______________ **3.** Hanger wires used for framing are No. _________ gauge galvanized wire.

______________ **4.** A(n) _________ is an area of the ceiling that is constructed at a finished height lower than the main ceiling.

A. fascia
B. soffit
C. loft
D. ridge

______________ **5.** A(n) _________ is a recessed area in the ceiling which provides a concealed location for the installation of light fixtures used for indirect lighting.

A. tray
B. soffit
C. light trough
D. bulb bay

______________ **6.** A(n) _________ is a rated enclosure which encloses elevators, air ducts, plumbing pipes, electrical wires, or other items which pass through the floors of a high-rise building.

A. shaftwell
B. shaftwall
C. either A or B
D. neither A nor B

______________ **7.** A(n) _________ partition is a partition designed to be assembled, disassembled, and reassembled in another location with minimal damage to the partition components.

_______________ **8.** Noncombustible _________-gauge metal studs are used for partition framing.

_______________ **9.** Noncombustible _________-gauge metal studs are used for load-bearing and nonload-bearing exterior walls.

_______________ **10.** A(n) _________ is used to mark the locations of partition tracks on the floor.

_______________ **11.** Light-gauge metal studs should be cut approximately _________″ shorter than the floor-to-ceiling dimension.

_______________ **12.** Heavy-gauge metal studs are securely anchored to the top and bottom tracks with _________.

A. welds
B. self-drilling, self-tapping framing screws
C. either A or B
D. neither A nor B

T F **13.** Multiple layers of drywall in shaftwalls must be offset at least 6″.

T F **14.** Hanger wires are available in lengths of from 2′ to 24′.

T F **15.** All demountable partitions, regardless of the manufacturer, are designed to divide space.

Soffits

_______________ **1.** Ceiling framing

_______________ **2.** Drywall

_______________ **3.** Main ceiling

_______________ **4.** Cabinet

_______________ **5.** Metal stud

_______________ **6.** Furring channel

_______________ **7.** Corner bead

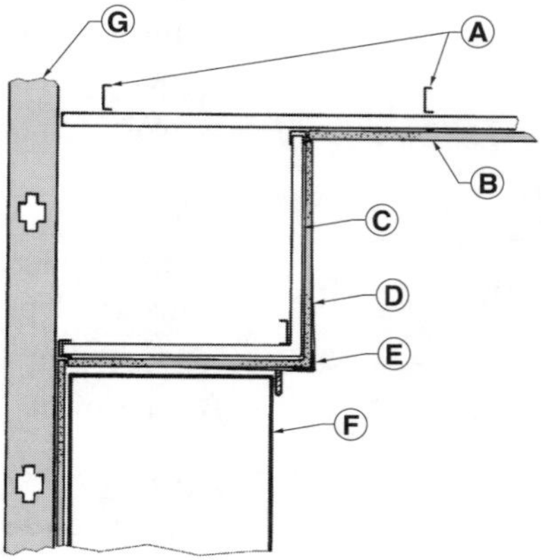

Clips

__________ **1.** Furring channel

__________ **2.** Screw

__________ **3.** Drywall

__________ **4.** Carrier channel

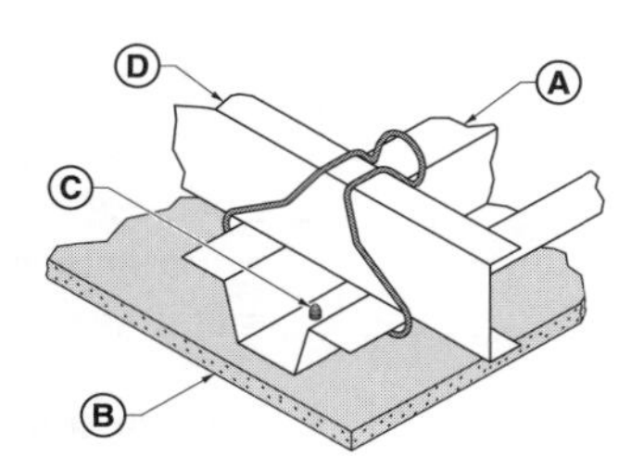

Hanger Wires

__________ **1.** Hanger wire

__________ **2.** Carrier channel

__________ **3.** Furring channel

__________ **4.** Wire tie

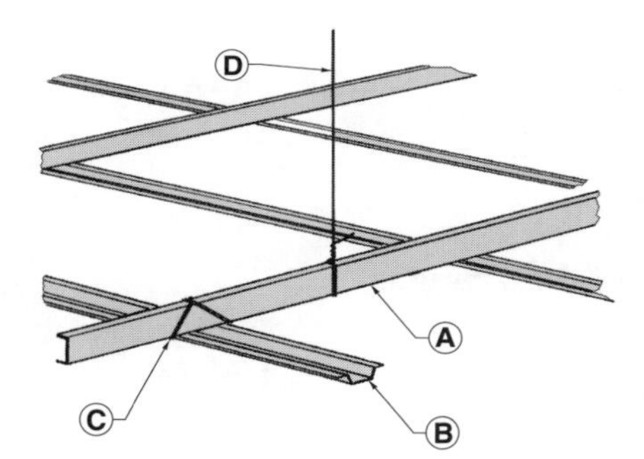

Demountable Partitions

__________ **1.** Flanged rail

__________ **2.** Gypsum panels

__________ **3.** Stud

__________ **4.** Floor track

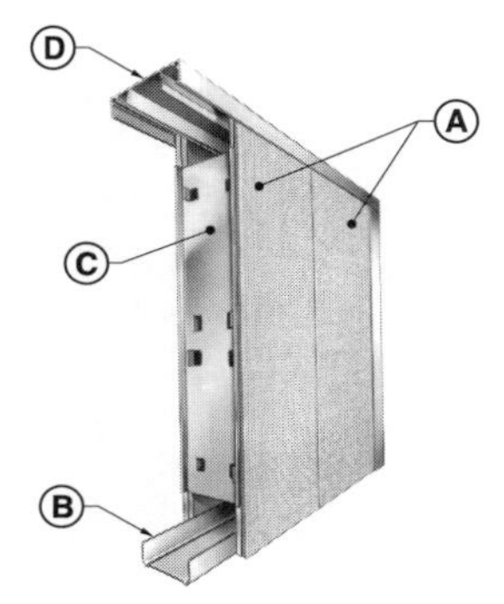

Stud Installation

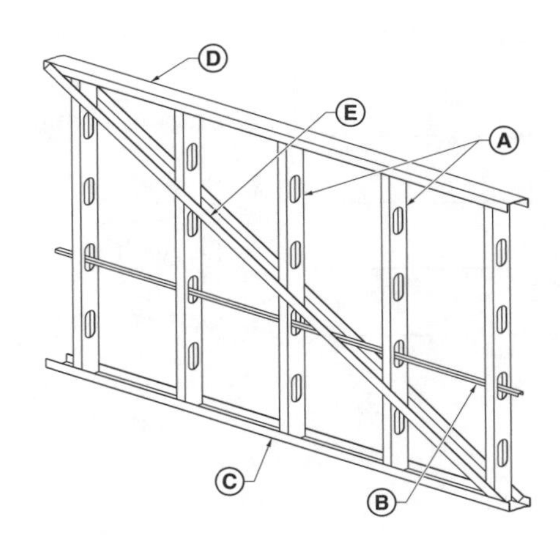

_____________ **1.** Top track

_____________ **2.** Load-bearing, heavy-gauge metal studs

_____________ **3.** Bottom track

_____________ **4.** Sway brace

_____________ **5.** Cold-rolled U-channel bracing

Light Troughs

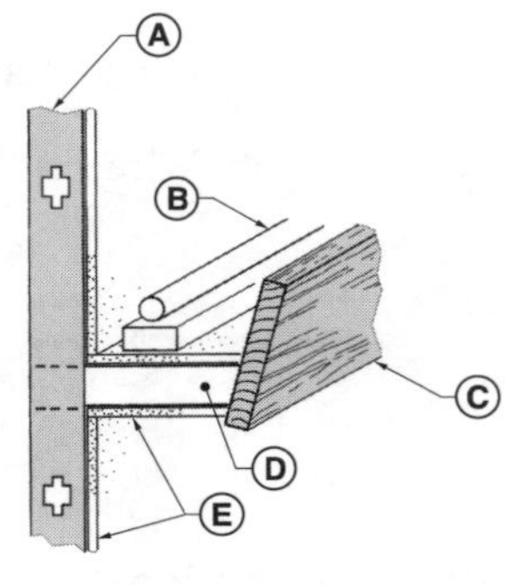

_____________ **1.** Drywall

_____________ **2.** Metal stud

_____________ **3.** Framing

_____________ **4.** Wood fascia

_____________ **5.** Light fixture

6 Sound- and Fire-Control Assemblies

Sound-control assemblies constructed from drywall sheets and wood or metal studs are used to control sound transmission through common walls. Fire-control assemblies constructed from multiple layers of drywall and wood or metal studs are used to provide fire protection.

SOUND-CONTROL ASSEMBLIES

The use of drywall assemblies for sound control has increased due to the emphasis placed on personal privacy in today's structures. Many times the occupants of office buildings, apartments, and condominiums are separated from their neighbors by common walls. A *common wall* is an interior wall shared by two or more occupancies. Common walls are designed to reduce the transmission of sound and fire from one occupancy to the other.

Sound-control assemblies are constructed from drywall sheets and wood or metal studs. A common method for controlling sound transmission consists of placing two rows of studs in the common wall. Double stud walls provide two surfaces completely isolated from each other which can absorb and transmit sound. See Figure 6-1.

The studs supporting the drywall sheets on one side of the double stud wall are not in contact with the drywall sheets on the opposite side. Sound waves striking the surface of the drywall sheets on one side of the double stud wall causes the drywall and studs on that side to vibrate. Unlike a single stud wall, the vibrations created by the sound waves are not directly transferred to the other side. The air space between the two wall surfaces allows the vibrations to dissipate, thereby reducing the force of the vibrations on the opposite side. As the sound transmitted to the opposite side is reduced or eliminated, the assembly serves as an effective sound barrier. See Figure 6-2.

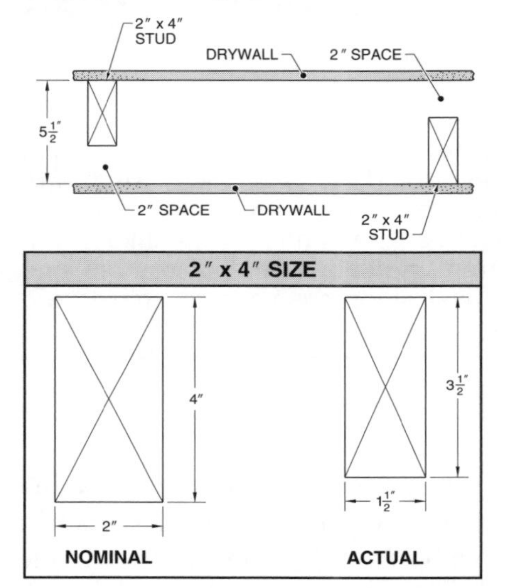

Figure 6-1. Double stud walls provide two surfaces completely isolated from each other which can absorb and transmit sound.

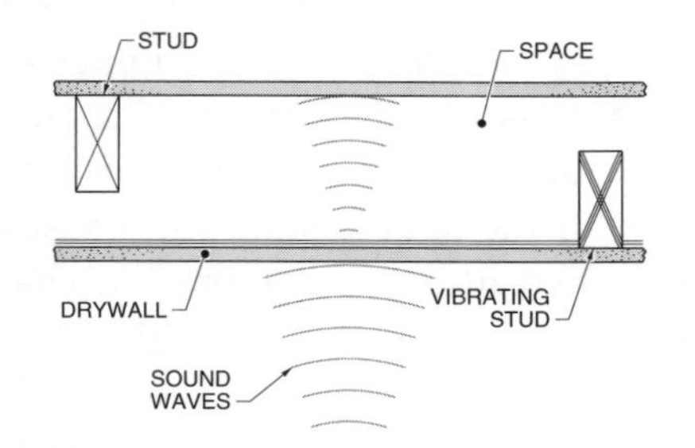

Figure 6-2. Sound waves striking the surface of the drywall sheet on one side of the double stud wall cause the drywall and studs on that side to vibrate.

To increase the sound control provided by a sound-control assembly, additional layers of drywall are applied to the framing members on each side of the wall. A sound-control assembly may also require the use of multiple layers of drywall which are different thicknesses. See Figure 6-3. This method is based on the assumption that multiple masses vibrate at different frequencies. These different frequencies tend to cancel the vibrations of each other. This concept is effective in reducing the amount of sound that is transmitted through the wall.

Sound may also be controlled to a lesser degree by installing sound insulation batts between the studs. See Figure 6-4. The use of sound insulation batts reduces only the sounds transmitted through the cavity of the wall. Sound transmission caused by the vibration of the entire wall assembly may be reduced, but not eliminated, with the use of sound insulation batts.

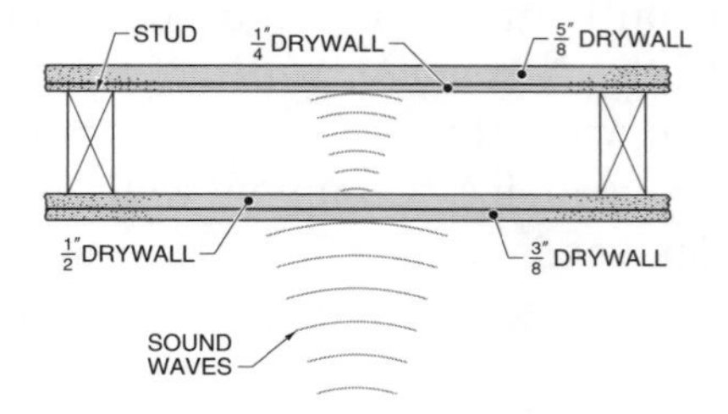

Figure 6-3. Multiple layers of drywall of different thicknesses reduce the amount of sound that is transmitted through the wall.

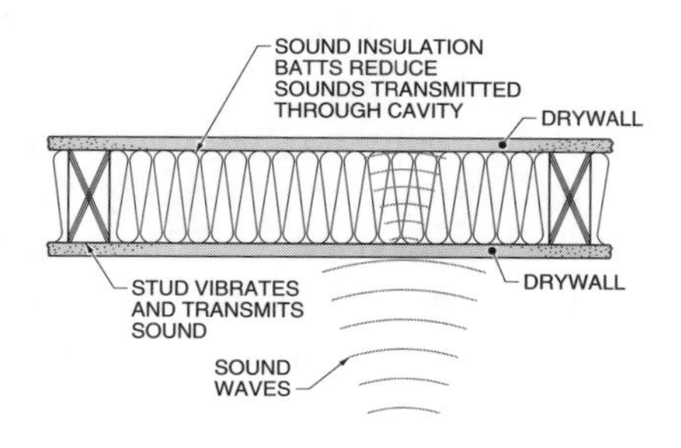

Figure 6-4. Sound insulation batts reduce only the sounds transmitted through the cavity of the wall.

Some sound-control assemblies employ a base layer of sound-deadening board installed beneath the drywall sheets to help reduce sound transmission. Because sound-deadening board is formed from wood or mineral fibers, it dampens the sound vibrations rather than transmitting them directly through the wall. A sound-deadening board may be placed on one or both sides of the wall. See Figure 6-5.

Resilient channels may be installed on the face of the framing members to reduce the transmission of sound. A *resilient channel* is a preformed metal channel which maintains an air space between the framing member and the drywall sheet. These specially-shaped channels are installed at right angles to the framing members at intervals indicated on the plans or in the specifications. See Figure 6-6. The drywall sheets are applied to the resilient channels with drywall screws in the same manner as it is applied to metal studs and drywall furring channels.

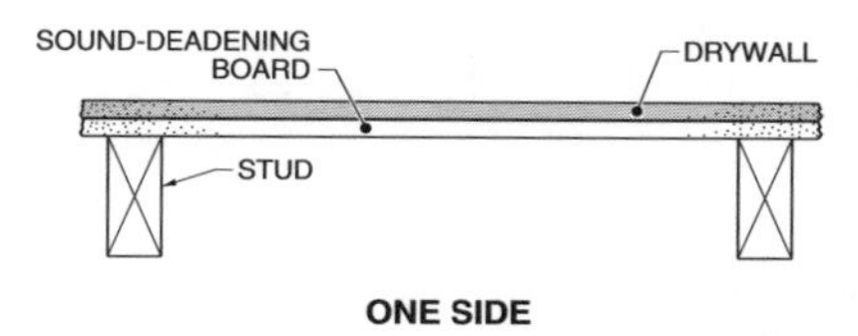

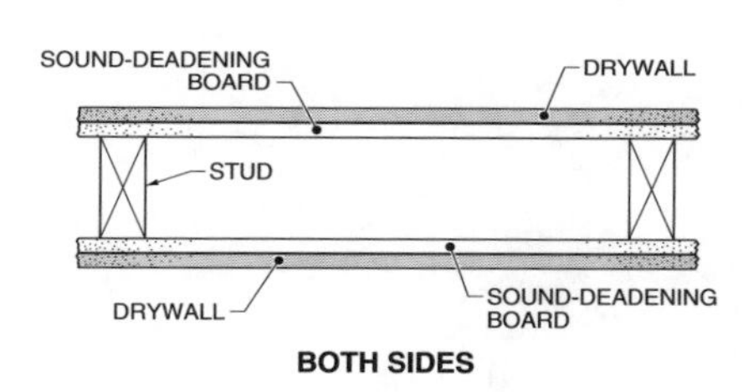

Figure 6-5. A sound-deadening board is installed beneath the drywall sheets on one or both sides of the wall.

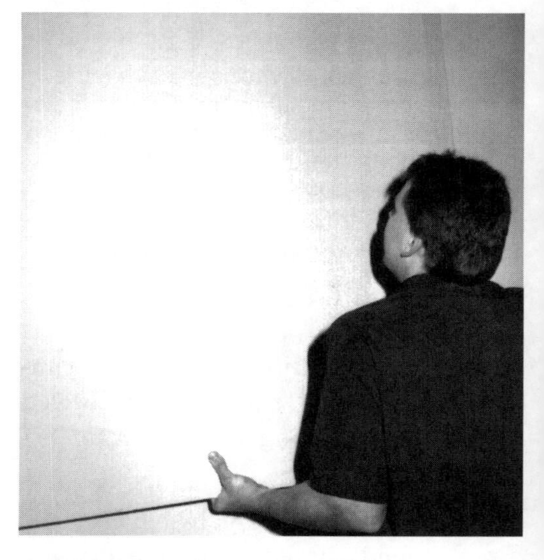

Different thicknesses of drywall may be used to create sound-control assemblies.

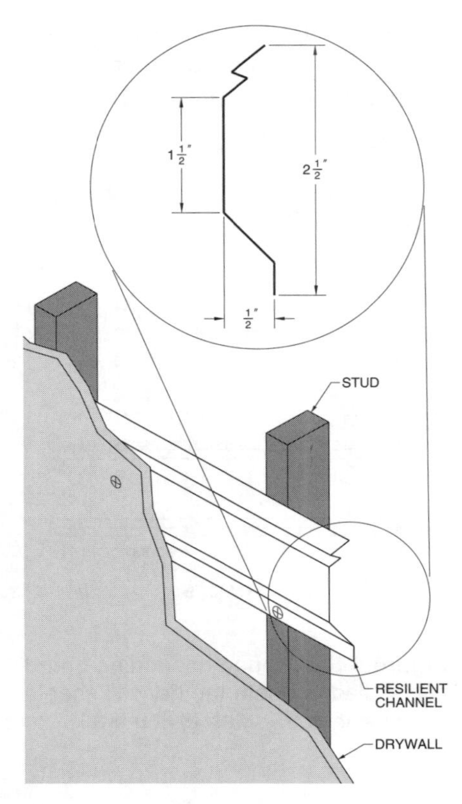

Figure 6-6. A resilient channel is a preformed metal channel which maintains an air space between the framing member and the drywall sheets.

Installing Sound-Control Assemblies

Five installation steps are recommended and should be followed when installing sound-control assemblies:

1. Do not leave gaps between the ends or edges of the drywall sheets or between the drywall sheets and other materials they butt against.
2. When constructing multilayer assemblies, the underlayer or layers must be installed with the same attention to detail as the finish layers. Voids in the underlayers reduce the sound control effectiveness of the assembly.
3. All cracks between walls, floor, ceiling or overhead structure, and abutting walls must be caulked with an approved sound-rated sealant.
4. Cutouts for electrical, plumbing, and mechanical installations should not be straight across from one side of the wall to the other. Such cutouts should be offset 12″ or one stud cavity.
5. When installing sound-control assemblies, do not cut corners. Sound-control assemblies must be constructed as shown on the plans and described in the specifications or according to local building code requirements.

INTERIOR FIRE-CONTROL ASSEMBLIES

Fire-control assemblies are constructed from multiple layers of drywall (coreboard) and wood or metal studs. They are designed to protect the building structure and the occupants from damage or injury due to fire. Fire-control assemblies contain the heat of a fire and control its rate of spread for a specified length of time. Fire-control assemblies include column and beam coverings, shaftwalls, multiple-layer applications, and tunnel corridors.

Special fire-control protection is required when steel columns and beams are used to construct the skeleton of the building. If not properly protected, steel members may collapse when exposed to the heat generated by a fire. Public corridors, stairwells, and other areas used by persons exiting the building must also be protected. The fire-control assemblies used in these locations provide protection for the occupants to safely exit the building in case of fire.

The length of time the structural members or the public areas must be protected is based on the type of assembly and is a product of the number and type of drywall layers used. Fire protection ratings range from a minimum of 45 min for demountable partitions to up to 4 hours for shaftwall assemblies. A *fire protection rating* is the length of time an assembly remains intact when exposed to fire. Typical fire protection ratings for various use groups are:

- A-1 Assembly, Theater – 3
- B Business – 2
- R-1 Residential, Hotels – 2
- R-3 Residential, One- and Two-Family Dwellings – 1

A one-hour fire-rated corridor is constructed by applying one layer of 5⁄8″ fire-shield type X drywall to each side of the framing members.

Corridors

Building codes require that all public corridors be constructed with fire-rated assemblies which provide one- or two-hour protection. See Figure 6-7. Corridors rated at one hour may be constructed with one-hour assemblies incorporating wood or metal studs and one layer of 5⁄8″ type X drywall applied to each side of the framing. This assembly usually requires that the drywall sheets be applied continuously from the floor to the structural members above unless the corridor has a fire-rated ceiling.

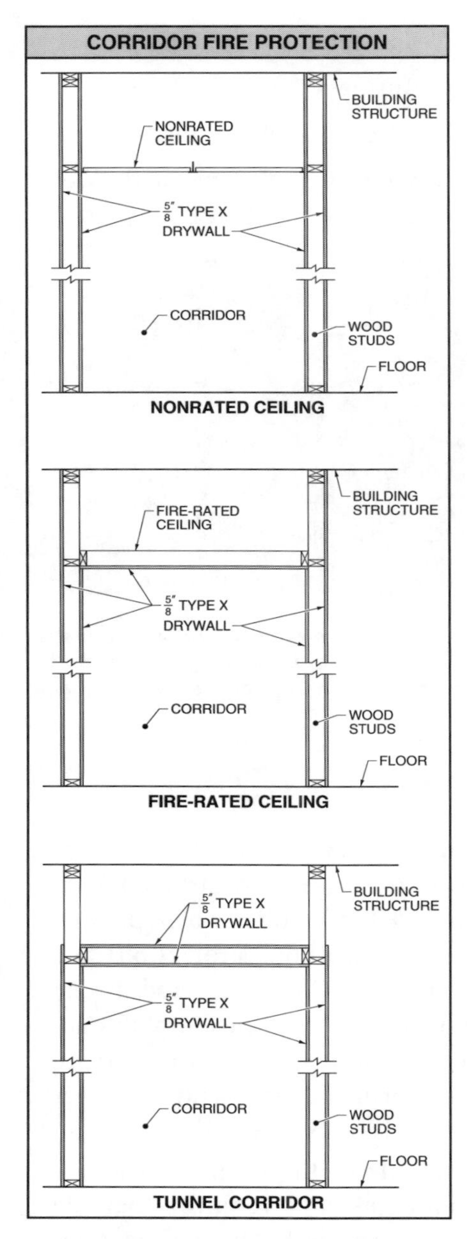

Figure 6-7. Building codes require that all public corridors be constructed with fire-rated assemblies which provide one- or two-hour protection.

When a fire-rated ceiling is constructed as part of the corridor, the drywall sheets applied to the corridor side of the framing may not be required to continue above the ceiling.

An alternative fire-rated assembly may be constructed in the form of a tunnel corridor. This assembly completely encloses the corridor with one or two layers of drywall applied on the wood or metal framing members. This assembly is commonly used when installation of air ducts, pipes, and conduits is required in the space above the corridor ceiling.

Stairways

Stairway walls may be constructed using shaftwall assemblies, wood or metal studs, and one or two layers of fire-resistant drywall. The drywall sheets applied on the stairway side of the walls must be continuous from the floor of the stairwell to the roof or overhead structure. Only the stair stringers may interrupt the continuous drywall application when they are installed within the finished plane of the wall. Should this condition occur, it may be necessary to modify the wall assembly to compensate for the break in the drywall at the stair stringers. Consult the plans and specifications for the specific treatment of this condition before proceeding.

Elevator Shafts

Elevator shafts constructed in wood-framed structures use multiple layers of fire-resistant drywall attached to both sides of the wood framing members to form a fire-rated shaft. Multiple-layer drywall applications require that the joints of various layers of drywall be offset at least 6″. Each layer is fastened to the framing members as required by the local building code. After the drywall sheets on the inside of the shaft are installed, the joints are taped and the heads of the fasteners are coated with joint compound.

The drywall sheets inside the elevator shaft are installed from the top down whenever possible. It is not advisable to work from the bottom up when installing drywall sheets on the inside of wood-framed elevator shafts due to the difficulty of setting up the scaffolding. This downward progression provides access for setting up the scaffolding through the open framing below the area where the work is being performed. Eight steps which should be followed to ensure the successful completion of the drywall inside an elevator shaft on a wood-framed structure are:

1. Set up the scaffolding approximately 8′-2″ below the elevator shaft ceiling.
2. Install the ceiling drywall.
3. When multiple layers of drywall are required, a 6″ or 12″ wide drywall rip is installed immediately below the ceiling drywall.
4. The first full drywall sheets of the first layer are installed immediately below the rip.
5. The top sheets of the second layer are installed. These sheets are butted against the ceiling drywall sheets.
6. The second full sheets of the first layer are installed. These must be notched around the scaffold planks if they extend through the wall framing of the shaft.
7. The second sheets of the second layer are installed.
8. Joint tape is applied to all joints and the fasteners are coated with joint compound.

This process is repeated as each level is completed. The notches cut in the first layer sheets must be filled after the scaffold has been moved to the next level. The sheet ends of each layer are reversed at the corners to ensure that the assembly provides the proper fire-resistance rating. See Figure 6-8.

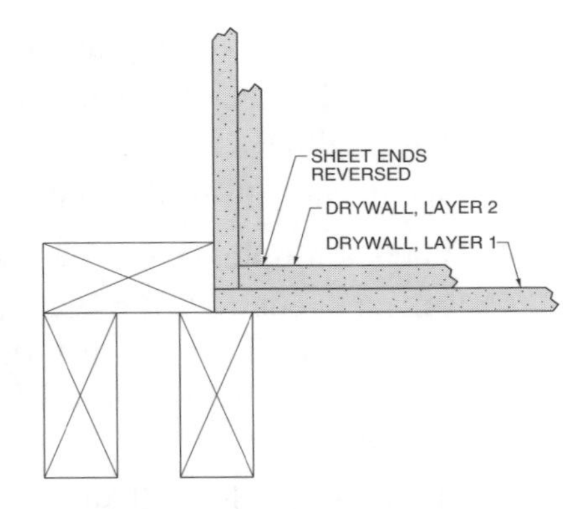

Figure 6-8. The sheet ends of each layer are reversed at the corners to provide the proper fire-resistance rating.

Firestop sealant is used to produce a fire-resistant barrier at cutouts for electrical, plumbing, and mechanical installations.

Structural Steel Beams and Columns

Structural steel beams and columns may be protected with a covering of two, three, or more layers of fire-resistant drywall. See Figure 6-9. The drywall may be attached directly to the steel members or to metal studs attached to the beams and columns. The total number of drywall layers installed depends on the degree of fire protection required. For example, two-hour fire protection generally requires two layers, three-hour fire protection requires three layers, etc.

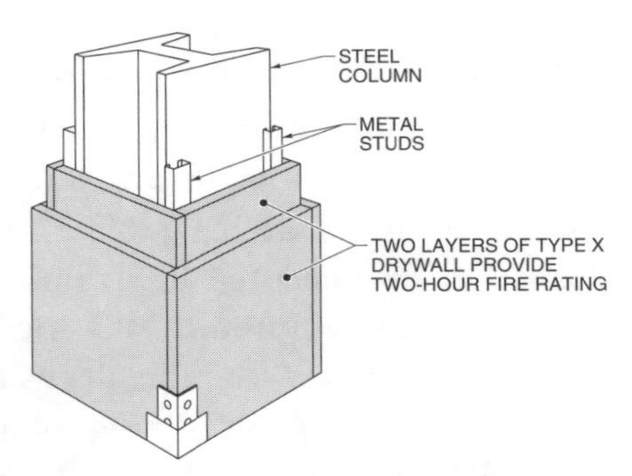

Figure 6-9. Structural steel beams and columns may be protected with a covering of two, three, or more layers of fire-resistant drywall.

Garage Walls

When the garage of any building shares a common wall with the space occupied by the residents or tenants, a fire-resistant assembly is required to separate the two areas. See Figure 6-10. Because garages are considered to be potentially hazardous areas due to the combustible materials associated with motor vehicles, fire-resistant drywall is used to construct fire-rated separations. The building design dictates the type and location of fire-rated separations.

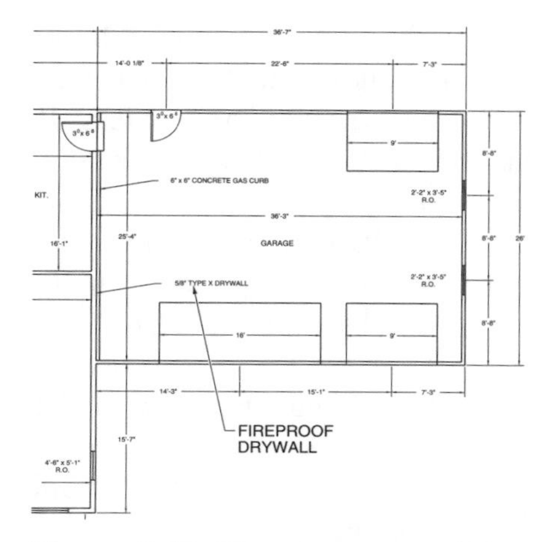

Figure 6-10. The common wall between a garage and the occupied space of a dwelling is required to be a fire-resistant assembly.

Fire-Resistant Chases. Locations in walls and floors where cutouts must be made to provide chases for electrical, plumbing, and mechanical installations must be protected with fire-resistant drywall. When wood-framed wall and floor members are used to form chases which are used for mechanical installations, the wood members must be protected with fire-resistant drywall inside and outside of the chase. When the chase is constructed in a wood floor assembly, the subfloor and floor joists supporting it must be protected.

When small chases are formed within the cavity formed by the ceiling joists, the subfloor and sides of the cavity must be covered with fire-resistant drywall. The drywall applied to the ceiling may provide the final side to the chase enclosure. See Figure 6-11.

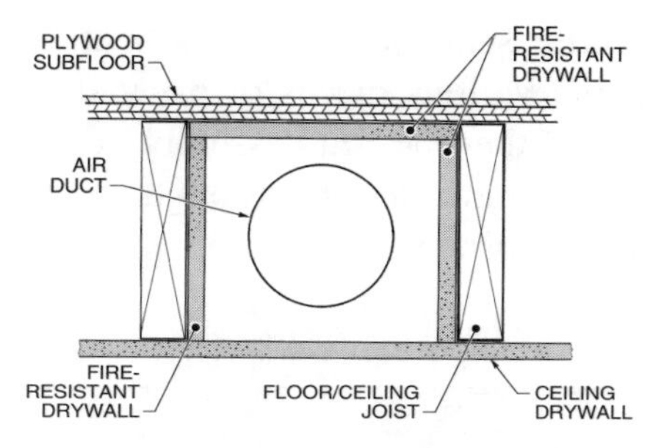

Figure 6-11. Chases formed by cavities between ceiling joists must be covered with fire-resistant drywall sheets.

Chases installed in wood stud walls require that fire-resistant drywall be applied to the insides of the studs. See Figure 6-12. In these applications, the drywall sheets used for covering the wall framing provide the remaining two sides of the enclosure. All fire-rated enclosures must be taped with joint tape and joint compound before they are closed up completely.

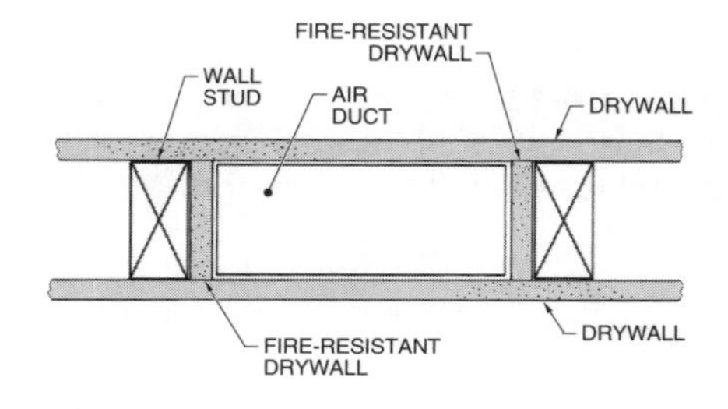

Figure 6-12. Chases formed by cavities between studs must be covered with fire-resistant drywall sheets.

Chases may require two or three layers of fire-resistant drywall to achieve the required fire protection. See Figure 6-13. These installations require sequencing the application of the various pieces of drywall to ensure that the final assembly does not contain straight-through joints at the corners.

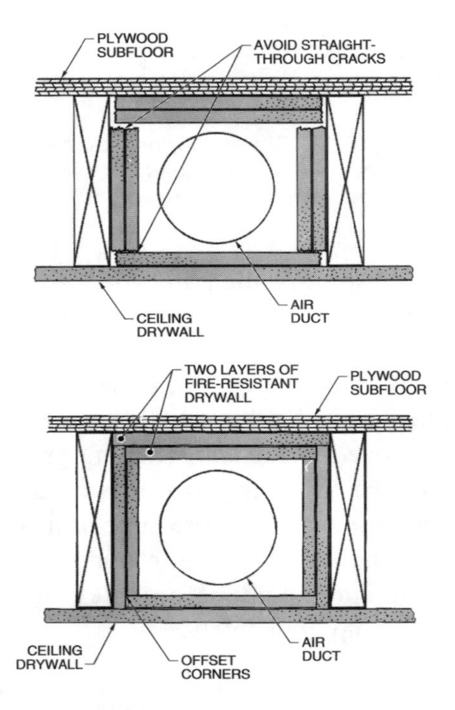

Figure 6-13. Chases may require two or three layers of fire-resistant drywall to achieve the required fire protection.

Draftstops

A *draftstop* is the divider or partition in the attic of a structure which retards the spread of fire and smoke within the building. Most local building codes limit the open area of an attic to 3000 sq ft or less. If the size of the building causes this limit to be exceeded, the attic is divided into 3000 sq ft sections using draftstops. Draftstops are usually shown on the building plans. See Figure 6-14.

Draftstops may be constructed with either wood or metal framing members. They must extend from the ceiling to the roof sheathing. The drywall sheets are installed on the draftstop framing using either nails or screws.

When installing drywall sheets on draftstops, the sheets must be accurately measured and cut to ensure a precise fit. Gaps between the drywall sheets, framing members, or pipes that penetrate the draftstop are unacceptable. Some draftstop installations require that the joints be covered with joint tape and coated with joint compound to ensure that the separation is as airtight as possible.

Draftstop installations usually require two drywall hangers. One method for completing the drywall installation places one drywall hanger on a ladder or scaffold and one drywall hanger on the floor beside the stack of drywall sheets. The worker on the ladder or scaffold makes the measurements and attaches the sheets to the framing after they have been cut by the other worker.

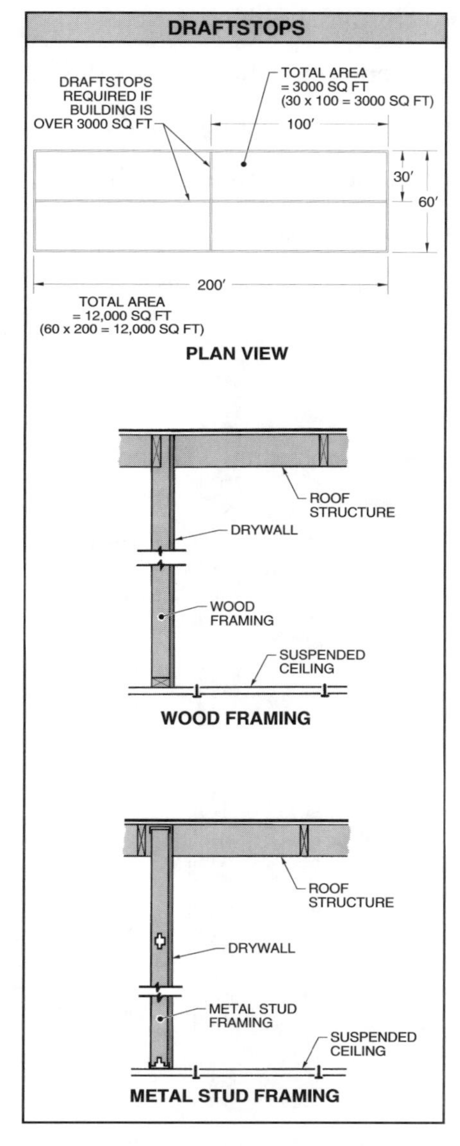

Figure 6-14. A draftstop is the divider or partition in the attic of a structure which retards the spread of fire and smoke within the building.

A second method also requires the use of two drywall hangers and a rolling scaffold. A small quantity of drywall sheets, sufficient to cover the area that can be reached from one scaffold set, is loaded on the scaffold platform. Both drywall hangers work together cutting the drywall sheets to size, installing them in place, and applying the joint tape. When the area is completed, the scaffold is moved and the procedure is repeated until the full length of the draftstop has been covered.

Interior Fire-Control Assembly Installations

Five installation steps are recommended and should be followed when installing fire-control assemblies:

1. Do not leave gaps between the ends or edges of the drywall sheets or between the drywall sheets and other materials they butt against.
2. When constructing multilayer assemblies, the underlayers must be installed with the same attention to detail as the finish layers. Voids in the underlayers reduce the fire-resistance effectiveness of the assembly.

3. All cracks and openings between the walls, floor, ceiling or overhead structure, and abutting walls must be caulked with an approved fire-rated sealant.
4. Cutouts for electrical, plumbing, and mechanical installations should be sealed with an approved fire-rated sealant.
5. Do not cut corners when installing fire-control assemblies. Fire-control assemblies must be constructed as shown on the plans and described in the specifications or according to local building code requirements.

EXTERIOR FIRE-CONTROL ASSEMBLIES

Gypsum panels can be used to construct fire-control assemblies on the exterior of a building. See Figure 6-15. The uses of gypsum panels, with its water-resistant core and face paper, continue to increase as additional exterior applications are devised. One example of these exterior applications is gypsum sheathing used as an underlayment for EIFS, cement plaster, masonry, and wood or metal siding applications. *Exterior insulation and finish systems (EIFS)* are exterior panel systems composed of sheathing, insulation board, and a finish. See Figure 6-16. The EIFS components from the wall out include:

- Gypsum sheathing
- Exterior insulation board adhesive
- Exterior insulation board
- Exterior reinforcing mesh
- Exterior basecoat
- Exterior textured finish

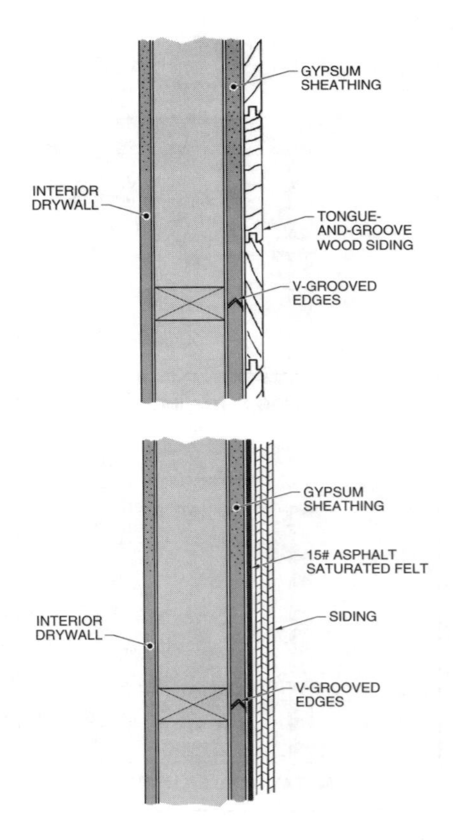

Figure 6-15. Drywall sheets can be used to construct fire-control assemblies on the exterior of a building.

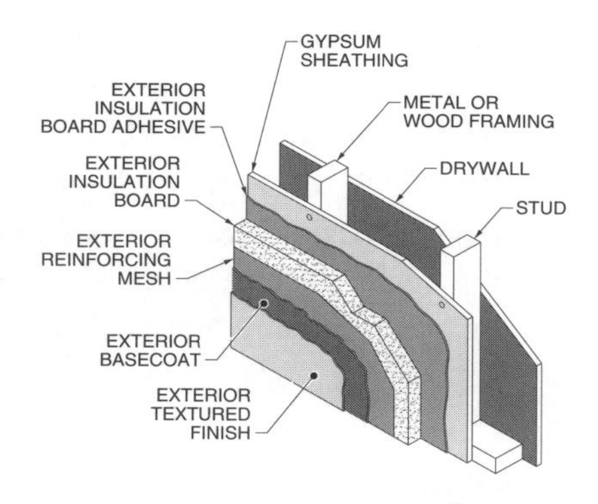

Figure 6-16. Exterior insulation and finish systems incorporate insulation board and polymer-based materials into an exterior finish.

Gypsum sheathing provides a fire-resistant base for a variety of exterior finishes. Horizontal surfaces of exterior ceilings and soffits may be covered with one or more layers of fire-resistant drywall.

Gypsum sheathing is designed to be applied under EIFS, cement plaster, masonry, and wood or metal siding. Regular drywall may also be used under wood siding but requires a covering of 15# asphalt-saturated felt under the wood siding.

Gypsum sheathing is manufactured with a special water-resistant core, face, and back paper. It is available in sheets ½″ and ⅝″ thick, 24″ and 48″ wide, and in lengths of 8′, 9′, and 10′. These sheets have either square or V tongue-and-groove edges. They provide an economical water and fire-resistant base layer which is ideal for the application of exterior finishes.

Gypsum sheathing is handled, measured, and cut in the same manner as regular drywall, except for sheets with tongue-and-groove edges. The tongue-and-groove edges must be carefully mated together as the sheets are installed. The gypsum sheathing is attached to framing members with nails or drywall screws at intervals as noted on the plans and in the specifications. The ends and edges of the sheets must be tight-fitting and all cutouts must be accurately made.

A covering of 15# asphalt-saturated felt is required when regular drywall is used as an exterior wallboard base.

Safety Precautions. Because gypsum sheathing is usually installed while the drywall hangers are working on a scaffold, these safety precautions should be observed:

1. Inspect the scaffold for any unsafe conditions.
2. When it is necessary to stock the gypsum sheathing on the scaffold, place only small quantities at each location. Overloading the scaffold may cause it to collapse.
3. Stack gypsum sheathing vertically on the scaffold. Secure the top of each stack to prevent the sheets from tipping.
4. Do not toss or drop scraps from the scaffold. Use a trash chute or other means for removing scrap.

United States Gypsum Company

United States Gypsum Company manufactures exterior products, systems, and assemblies that are lightweight and fire-resistant, and are suitable for commercial, industrial, and residential buildings.

Sound- and Fire-Control Assemblies

Trade Competency Test

Name Date

Assemblies

______________ **1.** _________ walls provide two surfaces completely isolated from each other which can absorb and transmit sound.

______________ **2.** A(n) _________ channel is a preformed metal channel which maintains an air space between the framing member and the drywall.

______________ **3.** A(n) _________ wall is an interior wall shared by two or more occupancies.

A. shared
B. common
C. shaftwall
D. exterior

T F **4.** A common method for controlling sound transmission consists of placing two rows of studs in a common wall.

______________ **5.** Sound transmission through a wall may be reduced somewhat by installing _________ batts between the studs.

______________ **6.** Studs supporting the drywall on one side of a(n) _________ wall are not in contact with the drywall on the opposite side.

______________ **7.** _________ channels may be installed on the face of the framing members to decrease the transmission of sound.

T F **8.** When installing sound-control assemblies, all cracks between walls, floor, ceiling or overhead structure, and abutting walls must be caulked with an approved sound-rated sealant.

______________ **9.** _________ walls are designed to reduce the transmission of sound and fire from one occupancy to another.

T F **10.** In a double stud wall, sound waves striking the surface of the drywall on one side are directly transferred to the other side.

T F **11.** Fire-control assemblies are constructed from multiple layers of drywall (coreboard) and wood or metal studs.

__________ **12.** Fire-control assemblies include _________.

A. tunnel corridors
B. column and beam coverings
C. shaftwalls
D. A, B, and C

T F **13.** Steel members may collapse when exposed to the heat generated by a fire if not properly protected.

__________ **14.** A fire protection _________ is the length of time an assembly will remain intact when exposed to fire.

T F **15.** Building codes require that all public corridors be constructed with fire-rated assemblies which provide one- or two-hour protection.

__________ **16.** Drywall sheets installed inside an elevator shaft should be installed _________.

A. from the bottom up
B. from the top down
C. more than three
D. A, B, and C

T F **17.** Public corridors, stairwells, and other areas used by persons exiting a building must be protected by fire-control assemblies.

__________ **18.** Structural steel beams and columns may be protected with a covering of _________ layers of fire-resistant drywall.

A. two
B. three
C. more than three
D. A, B, and C

T F **19.** A common wall separating a garage and a space occupied by building occupants does not require a fire-resistant assembly.

__________ **20.** A(n) _________ is a divider or partition in the attic of a structure which retards the spread of fire and smoke within the building.

EIFS

_____________ **1.** Exterior basecoat

_____________ **2.** Gypsum sheathing

_____________ **3.** Exterior reinforcing mesh

_____________ **4.** Gypsum panel

_____________ **5.** Exterior insulation board

_____________ **6.** Exterior textured finish

_____________ **7.** Metal or wood framing

_____________ **8.** Exterior insulation board adhesive

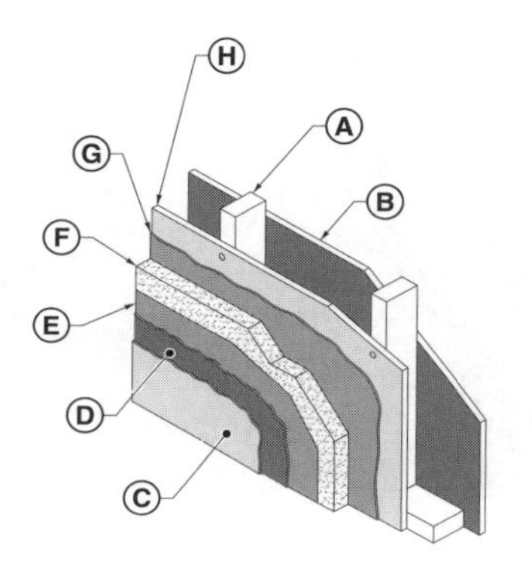

Draftstops

T F **1.** A draftstop is required at A.

T F **2.** A draftstop is required at B.

_____________ **3.** A minimum of _________ draftstops are required at C.

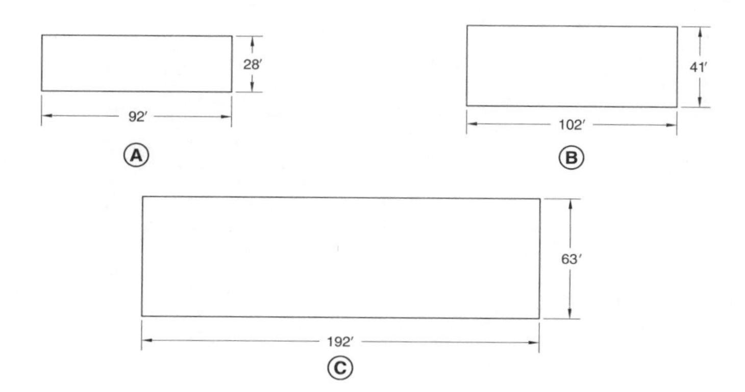

Fire-Resistant Chases

T F **1.** Duct chase A is properly fire-rated.

T F **2.** Duct chase B is properly fire-rated.

T F **3.** Duct chase C is properly fire-rated.

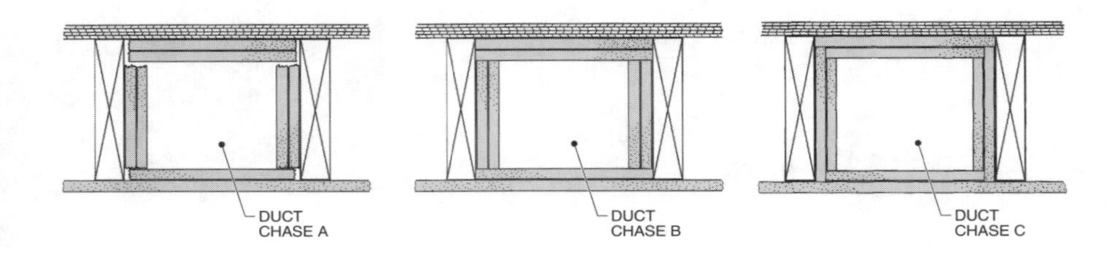

Fire Ratings

____________ **1.** Wall A has a(n) _________-hour fire rating.

____________ **2.** Column A has a(n) _________-hour fire rating.

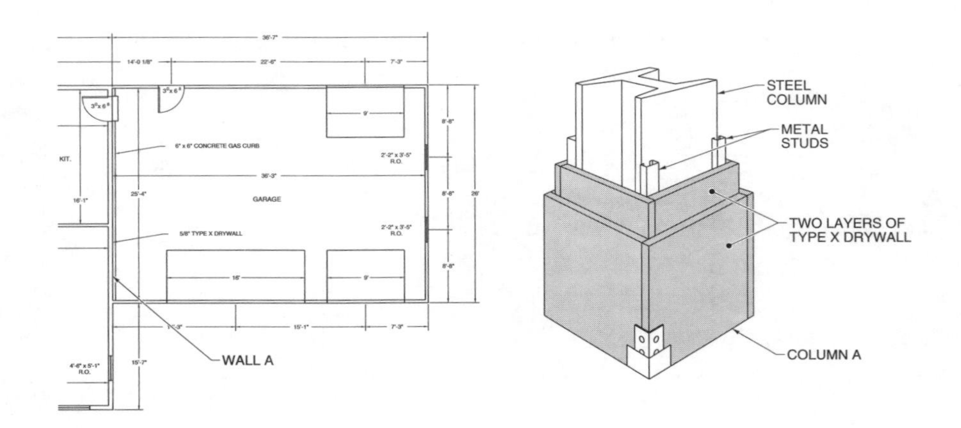

7 Special Installations and Materials

Special installations of drywall use moisture-resistant drywall, gypsum shaftwall liner, exterior ceiling and soffit drywall, and gypsum sheathing to provide protection from moisture and fire. Prefinished drywall is used in demountable partition systems or applied to existing walls. Drywall may be bent, curved, and cut at angles as required by the installation.

BENDS, CURVES, AND ANGLES

Drywall is commonly installed on flat surfaces such as partitions, walls, and ceilings. However, special applications often require that the drywall be bent, curved, or cut at an angle to complete the installation. Special procedures such as decreased framing member spacing, moistening, and backcutting may be required.

Bends

A bend is formed by shaping the surface of a drywall sheet. Bending drywall sheets requires the use of a special flexible drywall or multiple layers of ¼″, ⅜″, ½″, or ⅝″ drywall. Drywall sheets are installed on both metal and wood framing members to form curved surfaces. The dimension of the radius and/or the rating of the assembly determines the thickness or combination of thicknesses, and type of drywall required. *Radius* is the distance from the centerpoint to the outer edge of a circle.

Prior to the introduction of flexible sheets, the use of one, two, three, or more layers of ¼″, ⅜″, ½″, or ⅝″ drywall was the only means available for covering curved surfaces with drywall. The other alternative was metal lath and plaster. See Figure 7-1.

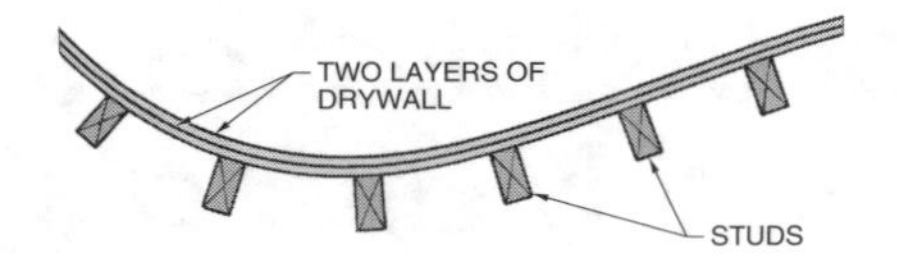

Figure 7-1. Multiple layers of drywall may be used to form a curved surface.

Curved surfaces with a radius of 8′ or more are constructed with the framing members located at the standard 16″ OC. However, as the dimension of the radius is reduced, the bend in the drywall becomes more severe and the framing members must be placed on 12″, 8″, 6″, 4″, or 2″ centers. The spacing of the framing members depends on the thickness, number of layers, and type of drywall being used. Drywall sheets applied on a curved surface with a small radius require more fasteners as the framing members become more closely spaced.

Curved drywall surfaces may be formed from metal framing members that can be easily fabricated to the required shape.

The drywall sheets may require moistening before installation. Moistening drywall sheets prior to installation on a curved surface allows the face paper to stretch as the sheet is bent. Once the drywall sheet has been fastened-in-place and has become thoroughly dry, the gypsum core of the drywall regains its original hardness and retains the curved shape. See Figure 7-2.

United States Gypsum Company

Figure 7-2. Curved surfaces may be covered with drywall by moistening the drywall sheets prior to installation.

Other applications, such as covering the jamb surface of an archway, require the drywall sheets to be bent in a very tight curve. To form a backcut, the drywall hanger cuts the back paper of the drywall sheet every ½″, ¾″, or 1″, depending on the dimension of the radius. This allows the sheet to break at these intervals and form the necessary curvature. A *backcut* is a series of relief cuts made on the back surface to facilitate bending the piece. See Figure 7-3. After the drywall sheet is in place, the flexible corner bead is installed and the joint compound is applied. *Flexible corner bead* is a corner bead with flanges that may be cut to allow the corner bead to be bent to any curved radius. This produces a smooth radius surface suitable for paint or other decoration.

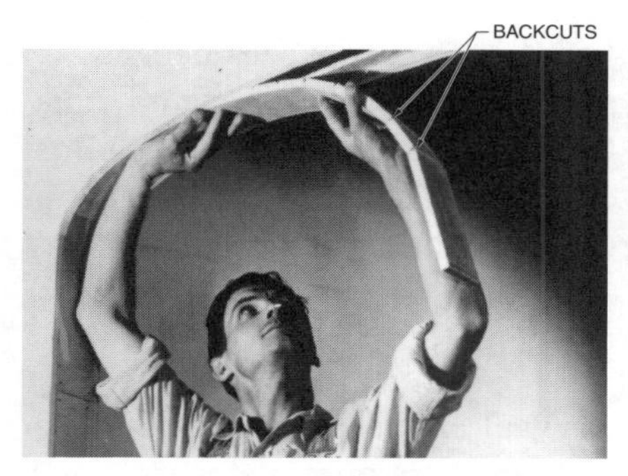

United States Gypsum Company

Figure 7-3. A backcut is a series of relief cuts made on the back surface to facilitate bending the piece.

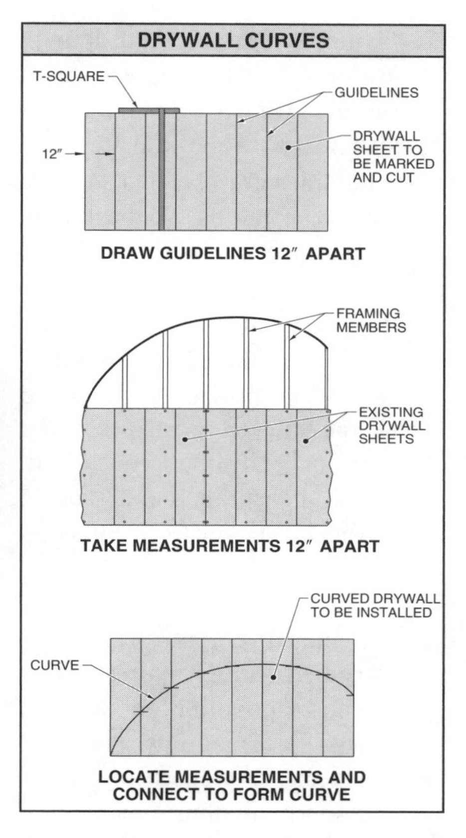

Figure 7-4. A curve is formed by cutting into an end or edge of a drywall sheet.

Curves

A curve is formed by cutting into an edge or end of a drywall sheet. Some drywall applications require the sheets on the ceiling to butt against a curved wall or soffit, or the sheets on the wall to butt against a curved ceiling. These installations are approached in a different manner than installations with straight-line conditions. Before the curved cut is made on the sheet to be applied to the ceiling or wall, the drywall sheet is laid out. See Figure 7-4.

The procedure for laying out the curve is:

1. Reference marks are placed along the side of the drywall sheet previously installed on the wall or ceiling at intervals of 12″ for a radius of 8′ or more, 6″ for a radius of 4′ to 8′, and 3″ for a radius less than 4′.

2. Using a T-square, guidelines are drawn at the same interval on the face of the drywall sheet. In the example, the guidelines are placed 12″ apart since the radius of the cut is approximately 10′.

3. Beginning at one end, the distance from the edge of the drywall sheet previously installed to the curved portion of the wall or ceiling is measured. These measurements are taken at the location of the intervals marked on the existing sheet.

4. All measurements are transferred to the drywall sheet at the corresponding location and marks are made where they cross the reference line. Each measurement must be marked on the drywall sheet at the corresponding reference line to produce an accurate curve. This process is continued until all dimensions are marked on the face of the drywall sheet.

5. When all dimensions have been made and marked on the sheet, a line is drawn from mark-to-mark to form a continuous curved line.

6. The drywall sheet is cut along the curved line.

The same steps are followed when making cuts at the end of the sheet unless the drywall sheet required for taking measurements has not been installed. If no drywall sheet is in place, the reference line interval marks are placed on the framing members. The successful application of drywall sheets adjacent to curved surfaces depends on precise measurements and cuts. Drywall sheets installed against curved surfaces are taped and finished in the normal manner.

90° Angles

Drywall applications on both metal and wood framing members which require cutting and fitting drywall sheets to form 90° angles are less complex than those involving curves. Forming 90° angles requires only straight lines and straight cuts to be made. See Figure 7-5. Whether installing drywall on ceilings which butt against angled walls or on walls which butt against angled ceilings, these installation steps are required:

1. The length of the sheet to be installed is measured along the edge of the sheet previously installed.
2. The width is measured at each end of the sheet at the widest (highest) point and at the narrowest (lowest) point.

3. Measurements are transferred to the drywall sheet being cut.
4. A chalk line is held between the points and snapped to mark the angle on the face of the sheet.
5. The drywall sheet is cut along this diagonal line.

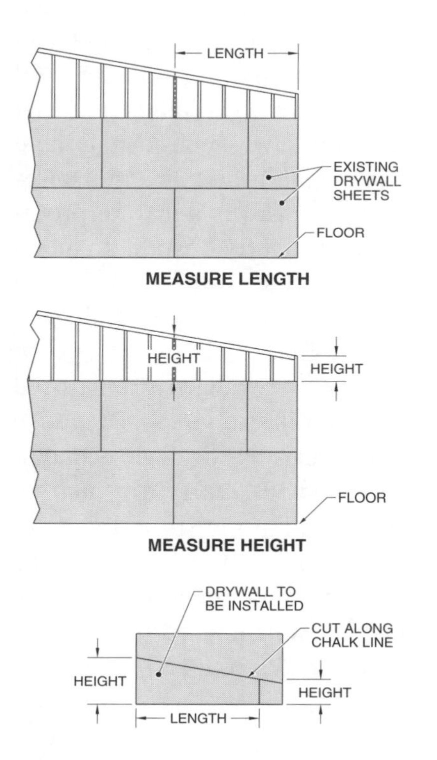

Figure 7-5. Forming 90° angles requires only straight lines and straight cuts to be made.

On projects with drywall sheets of sufficient length to cover one-half of the length of the wall or ceiling, measurements are taken in the center and at each end. See Figure 7-6. Both measurements are transferred to the drywall sheet, a chalk line is snapped between the marks, and the cut is made along the chalk line.

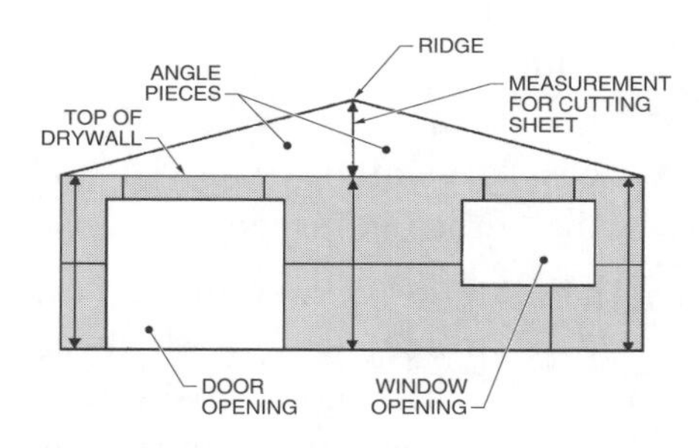

Figure 7-6. Measurements are taken in the center and at each end for drywall sheets which cover one-half of the length of the wall or ceiling.

All drywall sheets must be precisely measured and cut so the cut edges fit tightly against the angle surfaces and are free from unsightly gaps. The completed installation is taped and finished in the normal manner.

Off Angles

Wall, ceiling, and soffit installations also require drywall sheets to be applied at off angles. An *off angle* is an angle greater or less than 90°. See Figure 7-7. When installing drywall sheets at off angles, these installation steps are required:

1. The drywall sheets are installed on one surface with the edge or end of the sheet in line with the plane of the adjacent surface.
2. The drywall sheets on the adjacent surface are installed with the edges or ends extending 3″ past the corner. The excess drywall is removed and the corner is rasped flush.

The outside corner of the off angle is protected with flexbead corner trim. The taping and finishing of these areas are completed in the normal manner.

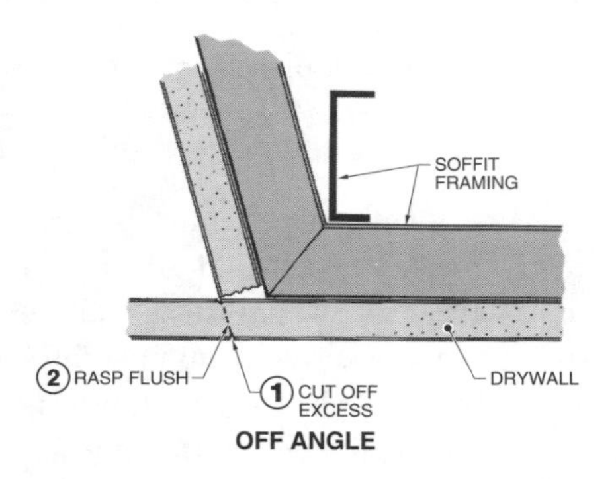

Figure 7-7. An off angle is an angle greater or less than 90°.

SPECIAL APPLICATIONS

Drywall is installed in a variety of applications to provide protection from moisture and fire. These applications include moisture-resistant drywall, gypsum shaftwall liner, and exterior ceiling and soffit drywall.

Moisture-Resistant Drywall

Moisture-resistant drywall is installed in areas where a high level of moisture may be present. These areas include bathrooms, restaurant kitchens, laundromats, etc. *Moisture-resistant drywall* is drywall in which the gypsum core contains asphalt and other materials which can withstand moisture and face and back paper which is treated to resist rot and mildew. Though it is able to withstand moisture, moisture-resistant drywall is not waterproof.

When properly installed, moisture-resistant drywall provides an economical and durable wall surface suitable for ceramic tile or other types of wall coverings used in high-moisture areas. Because moisture-resistant drywall is manufactured specifically for use in areas with high moisture, the deterioration of both core and face paper, a common occurrence when regular drywall is used in these areas, is greatly reduced or totally eliminated.

Moisture-resistant drywall is measured, cut, and attached to the metal or wood framing members in the same manner as regular drywall. However, the bottom sheets must be installed leaving a ¼″ space between the bottom edge of the sheet and the shower pan or bathtub. See Figure 7-8. This ¼″ space allows any moisture that migrates behind the tile or other covering to evaporate into the wall cavity rather than soak into the drywall core. All cut edges, sheet ends, and cut holes are coated with a special moisture-resistant sealant. This application of sealant prevents moisture saturation at these locations.

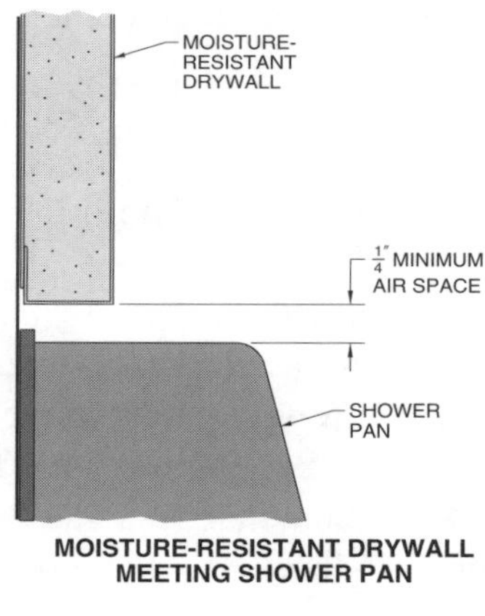

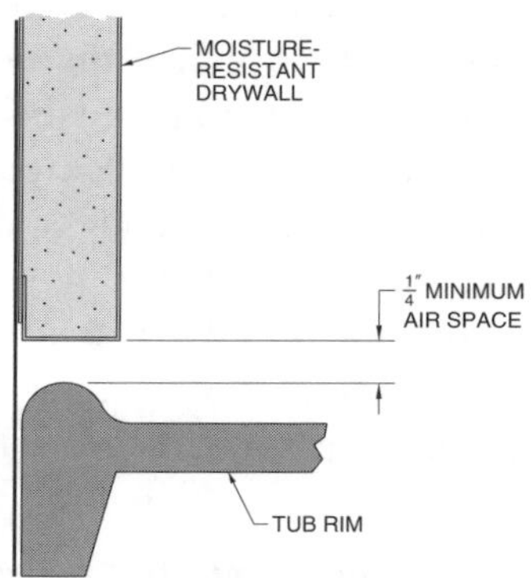

Figure 7-8. A ¼″ minimum air space must be left between the bottom edge of moisture-resistant drywall and the shower pan or bathtub.

United States Gypsum Company

DUROCK™ cement board is a multipurpose building panel that is water-damage resistant and provides a base for tub and shower tile areas.

Where moisture-resistant drywall is intended to be the finished wall or ceiling surface, it may be taped and finished similar to regular drywall. When ceramic tile or other coverings are to be installed on moisture-resistant drywall, taping and finishing is not normally required.

When ceramic tile is installed, the entire surface of the moisture-resistant drywall is coated with a waterproof mastic. After the mastic coating dries, the thin-set tile mastic is applied and the tile is installed and grouted. A nonhardening caulking may be applied at the point where the tile meets the shower pan or the bathtub. The ¼″ air space must not be filled with caulking. See Figure 7-9.

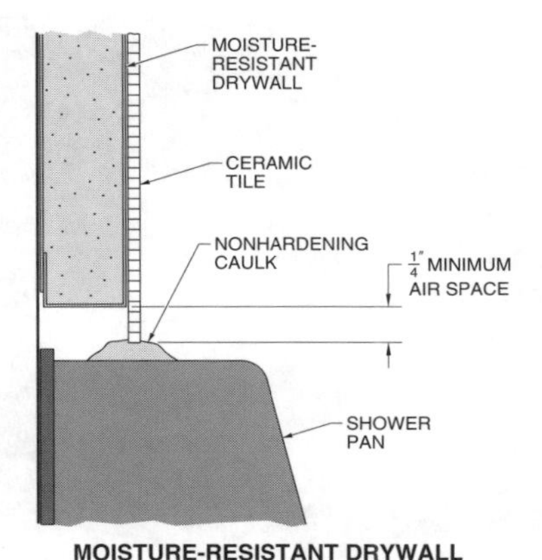

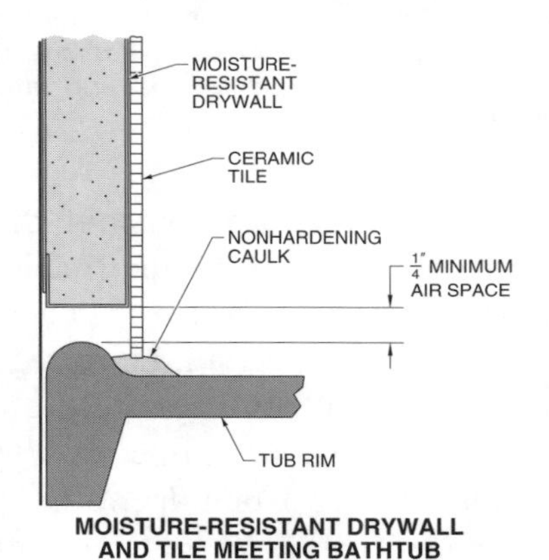

Figure 7-9. The ¼″ air space must not be filled with caulking when installing ceramic tile.

Moisture-resistant drywall can also be used as a backing material (base layer) for the redwood siding installed in sauna baths. When moisture-resistant drywall is used for this purpose, it is installed in the conventional manner before being covered with redwood siding.

Drywall applications in high-moisture areas where regular drywall is unsuitable may be completed using moisture-resistant drywall for walls and ceilings. In most of these instances, moisture-resistant drywall is installed in the same manner as regular drywall. Moisture-resistant drywall is available in sheets 48″ wide, ½″ and ⅝″ thick, and in standard lengths of 8′, 10′, and 12′.

Gypsum Shaftwall Liner

Gypsum shaftwall liner (coreboard) is a gypsum-based sheet material with either tongue-and-groove or square edges. See Figure 7-10. The core of gypsum shaftwall liner has properties similar to the core of moisture-resistant drywall. It contains materials which reduce the level of deterioration caused by moisture. The face and back paper is treated to resist rot and mildew and is colored. Although gypsum shaftwall liner may be exposed to moisture present in elevator and mechanical shafts, the deterioration of both core and paper is minimal.

Gypsum shaftwall liner sheets are measured and cut in the same way as regular drywall sheets. When sheets with tongue-and-groove edges are used, additional attention must be given when placing the sheets against those previously installed to ensure a proper fit. Square-edge sheets are inserted into the channel provided by the special shaftwall studs. After the entire shaft has been assembled, a cover sheet of standard drywall is applied.

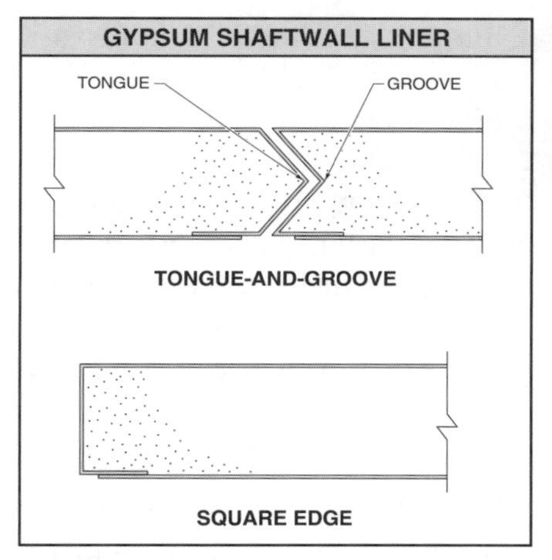

Figure 7-10. Gypsum shaftwall liner (coreboard) is a gypsum-based sheet material with either tongue-and-groove or square edges.

Gypsum shaftwall liner is designed primarily for interior applications as one component of an assembly. The framing parts and gypsum shaftwall liner must be supplied by the same manufacturer. Fire tests are conducted using shaftwall materials from the same manufacturer, so products from different manufacturers must not be combined. Doing so could cause the fire rating of the assembly to be nullified. Gypsum shaftwall liner is available in sheets 24″ wide, 1″ thick, and in standard lengths of 8′, 10′, 12′, 14′, and 16′.

Exterior drywall has a thicker paper and contains ingredients to protect the drywall from exposure to moisture.

Exterior Ceiling and Soffit Drywall

Exterior ceiling and soffit drywall is a moisture-resistant, gypsum-based sheet material with thicker face and back paper treated to resist mildew. It is installed on exterior ceilings and in soffit areas where speed, economy, and ease of installation is desired. See Figure 7-11.

The gypsum core of exterior ceiling and soffit drywall contains additional ingredients which reduce the level of deterioration that is caused by exposure to moisture. The added paper thickness helps reduce sagging of the sheets between ceiling support members and the treatment reduces the discoloration caused by the presence of mildew.

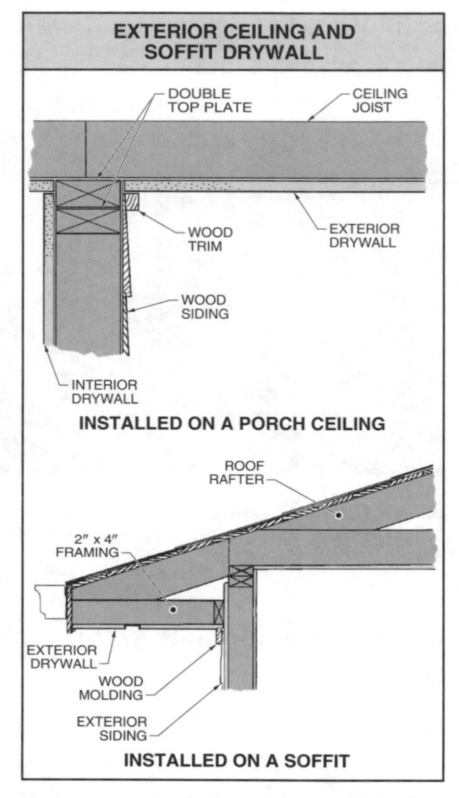

Figure 7-11. Exterior ceiling and soffit drywall is a moisture-resistant, gypsum-based sheet material with thicker face and back paper treated to resist mildew.

Exterior ceiling and soffit drywall has the appearance of regular drywall. It is measured, cut, and attached to the metal or wood framing members in the same manner as other drywall sheets. After it has been installed, exterior ceiling and soffit drywall is taped and finished with a special exterior ceiling and soffit finishing compound. When the area being covered exceeds 20′ in any direction, the exterior ceiling and soffit drywall sheets should be installed with the ends spaced ¼″ apart. This ¼″ space is provided to allow for the expansion and contraction of the sheets and is finished with a control joint drywall trim.

Exterior ceiling and soffit drywall is not suitable in areas where a high level of moisture is present for an extended period. These areas may be completed using lath and cement plaster, metal soffit panels, etc. In most instances, exterior ceiling and soffit drywall is installed in the same manner as regular drywall. Exterior ceiling and soffit drywall is available in sheets 48″ wide, ½″ and ⅝″ thick, and in standard lengths of 8′, 10′, and 12′.

Gypsum Sheathing

Gypsum sheathing is an exterior wallboard consisting of a water-repellent gypsum core with a water-repellent paper on face and back surfaces. It is typically installed to form a backing for a variety of exterior wall finishes. Edges may be tongue-and-groove for horizontal applications or square for vertical applications. When properly installed, gypsum sheathing provides an economical and durable wall surface suitable for covering with EIFS, lath and cement plaster, wood or metal siding, masonry veneer, etc. See Figure 7-12. Gypsum sheathing is manufactured primarily for exterior use.

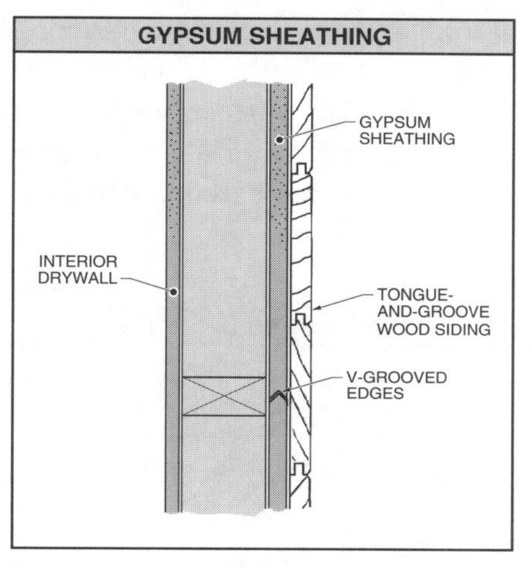

Figure 7-12. Gypsum sheathing is a gypsum-based sheet material with tongue-and-groove edges and is typically installed to form a backing for a variety of exterior wall finishes.

On most gypsum sheathing applications, the sheets are measured, cut, and attached to the metal or wood framing members the same as regular drywall sheets. The tongue-and-groove edges require additional attention when placing the sheets against those previously installed to ensure that they properly fit together. After the gypsum sheathing has been installed and the attachments have been approved, the exterior covering is applied. A dry sheet of asphalt-saturated felt or sisal paper may be required when gypsum sheathing is used as a backing material for certain exterior finishes. See Figure 7-13. This dry sheet, with suitable laps at the sides and ends, is attached to the surface of the gypsum sheathing with staples after the fastener inspection and prior to applying the exterior finish.

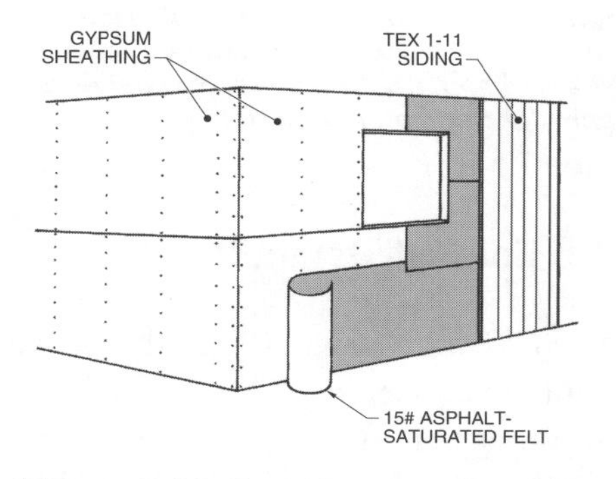

Figure 7-13. A dry sheet of asphalt-saturated felt or sisal paper may be required when gypsum sheathing is used as a backing material for certain exterior finishes.

There are interior drywall applications where regular drywall is unsuitable. These can often be completed using gypsum sheathing for walls and ceilings. In most instances, gypsum sheathing is installed in the same manner as regular drywall. Gypsum sheathing is available in sheets 24″ and 48″ wide, ½″ and ⅝″ thick, and in a standard length of 8′.

United States Gypsum Company

Gypsum sheathing, consisting of a fire-resistant core encased in water-repellent paper on both sides may be used for exterior steel or wood-framed construction on residential dwellings and light commercial buildings.

Prefinished Drywall

Prefinished drywall is drywall which is covered with vinyl, fabric, or other materials. This type of drywall sheet is used in demountable partition systems or applied to existing wall surfaces. Prefinished drywall requires no taping, painting, or other finishing. It is available in a wide variety of colors and textures. Prefinished drywall is available in sheets 48″ wide, ½″ and ⅝″ thick, and in a standard length of 8′. However, sheets may be ordered in other lengths and widths for special installations.

Prefinished drywall sheets are available with both square or chamfered edges. See Figure 7-14. Square-edge sheets are commonly used for demountable partition systems which use battens to cover the joints. Sheets with beveled edges may be used for installations which do not require battens. The chamfered edges provide a V-groove appearance when the edges of the two sheets are butted together. A *chamfer* is an angled cut that extends from the surface to the edge of a piece of material. A *bevel* is an angle cut that extends from surface-to-surface of a piece of material.

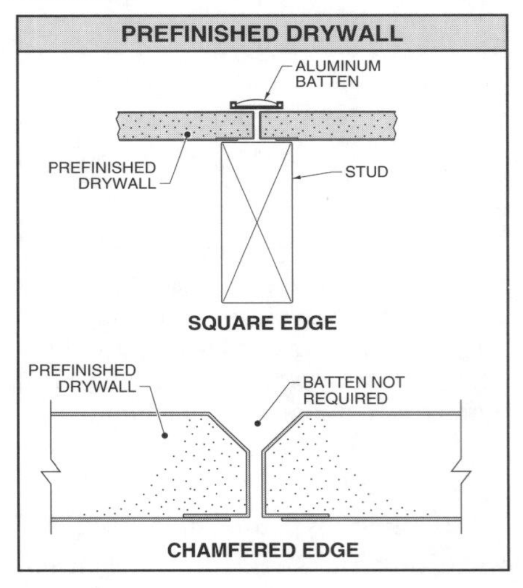

Figure 7-14. Prefinished drywall is drywall which is covered with vinyl, fabric, or other materials.

The face and edges of prefinished drywall must be protected to avoid damage during handling and installation. If the edges are damaged, a gap may result when the sheets are butted together. When the damage has occurred on one edge only, these sheets may be used for rips but must not be installed as full sheets. The gaps caused by damaged sheet edges are unacceptable, and if allowed to occur will adversely affect the appearance of the entire finished product. Cutouts for utility fixtures must be accurately measured and cut.

Panel adhesive may be used to install prefinished drywall sheets on furring strips, wood or metal framing members, and as a final layer on multilayer assemblies. When applied on wood framing members, a few color-matched nails may be used to hold the sheets in place until the adhesive sets and reaches its full bond strength. When applied on metal framing, or if the pattern of the prefinished drywall would create an undesirable effect by using color-matched nails, the sheets may be bowed or warped by laying them flat on the floor with the prefinished face up. A 4′ length of 4″ × 4″ is placed under each end of the stack for 24 hours. This method of prebowing causes the center of the sheet to place enough pressure against the face of the framing members or underlayer, when fastened at the top and bottom, to ensure a proper adhesive bond. Top-set base or wood base molding conceals the fasteners.

Prefinished drywall sheets may be installed over existing plaster or drywall walls or a drywall base layer using contact cement. In such installations, the contact cement is applied to both the wall surface being covered and the back side of the prefinished drywall sheet. The contact cement is allowed to dry and the sheet is installed. Each sheet must be positioned in its precise location before making contact with the substrate since they cannot be moved afterwards. Water-based contact cement is recommended for laminating prefinished drywall sheets as solvent-based contact cement may cause bubbles to develop in the vinyl covering.

Prefinished drywall sheets must be handled carefully and should be stored out of high-traffic areas. Dollies, forklifts, scaffolds, or other equipment may damage the edges of the sheets if allowed to strike them. Stacks of prefinished drywall sheets must not be used as workbenches. Enough blocking must be placed under the stacks to ensure the sheets are kept flat. Insufficient blocking may cause the sheets to warp or ripple.

United States Gypsum Company

Prefinished drywall is available in a variety of textures and colors.

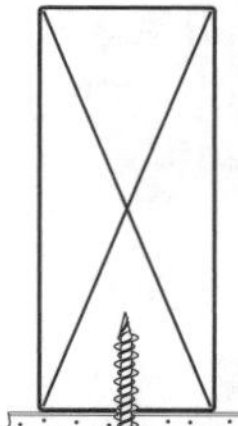

Special Installations and Materials

Trade Competency Test

Name Date

Special Installations

__________ **1.** A(n) _________ is formed by shaping the surface of a drywall sheet.

__________ **2.** A(n) _________ is formed by cutting into an edge or end of a drywall sheet.

__________ **3.** Only straight lines and straight cuts are required for drywall forming _________° angles.

T F **4.** Moisture-resistant drywall is waterproof.

__________ **5.** A(n) _________ angle is an angle greater or less than 90°.

T F **6.** The face paper of exterior ceiling and soffit drywall is the same thickness as the face paper of regular drywall.

T F **7.** The back paper of exterior ceiling and soffit drywall is moisture-resistant.

T F **8.** Gypsum sheathing has tongue-and-groove edges.

__________ **9.** _________ drywall is drywall which is covered with vinyl, fabric, or other materials.

__________ **10.** The _________ is the distance from the centerpoint to the outer edge of a circle.

A. diameter
B. radius
C. circumference
D. sector

______________ **11.** The bottom sheets of moisture-resistant drywall are installed with a(n) _________″ space between the bottom edge of the sheet and the shower pan or bathtub.

A. ¼ C. ¾
B. ½ D. neither A, B, nor C

T F **12.** Gypsum shaftwall liner is designed primarily for interior applications.

T F **13.** Exterior ceiling and soffit drywall is available in standard lengths of 6′, 8′, 10′, and 12′.

T F **14.** Gypsum sheathing is manufactured primarily for interior use.

______________ **15.** A(n) _________ is a series of relief cuts made on the back surface.

T F **16.** When ceramic tile is installed, the entire surface of the moisture-resistant drywall is coated with a waterproof mastic.

______________ **17.** Curved surfaces with a radius of _________′ or more are constructed with the framing members located at the standard 16″ OC.

______________ **18.** Moisture-resistant drywall is available in sheets _________″ wide.

______________ **19.** Gypsum shaftwall liner is available in sheets _________″ wide.

______________ **20.** When the area being covered with exterior ceiling and soffit drywall exceeds _________′ in any direction, the ends should be spaced ¼″ apart.

A. 10 C. 20
B. 12 D. 24

______________ **21.** Gypsum sheathing is available in sheets _________″ wide.

A. 24 C. either A or B
B. 48 D. neither A nor B

T F **22.** Prefinished drywall may be ordered in custom lengths for special installations.

_____________ **23.** A(n) _________ is an angled cut that extends from surface-to-surface of a piece of material.

_____________ **24.** A(n) _________ is an angled cut that extends from the surface to the edge of a piece of material.

T F **25.** Prefinished drywall sheets may be installed over existing plaster or drywall sheets using contact cement.

Exterior Ceiling and Soffit Drywall

_____________ **1.** Interior drywall

_____________ **2.** Exterior drywall

_____________ **3.** Wood trim

_____________ **4.** Wood siding

_____________ **5.** Ceiling joist

_____________ **6.** Double top plate

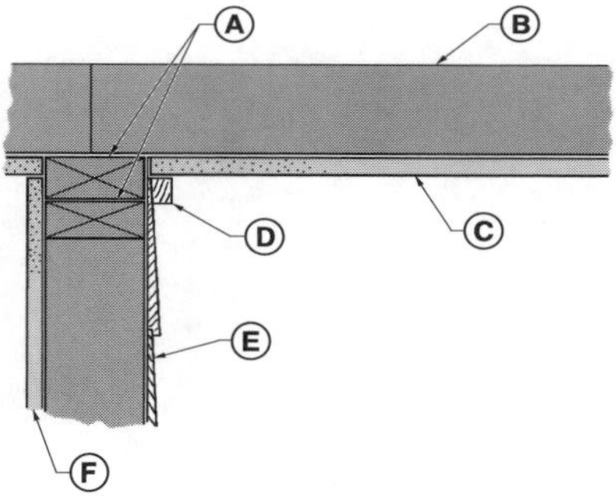

A B C D E F

Edges

_____________ **1.** Square-edge

_____________ **2.** Chamfered

_____________ **3.** Tongue-and-groove (gypsum)

_____________ **4.** Tongue-and-groove (wood)

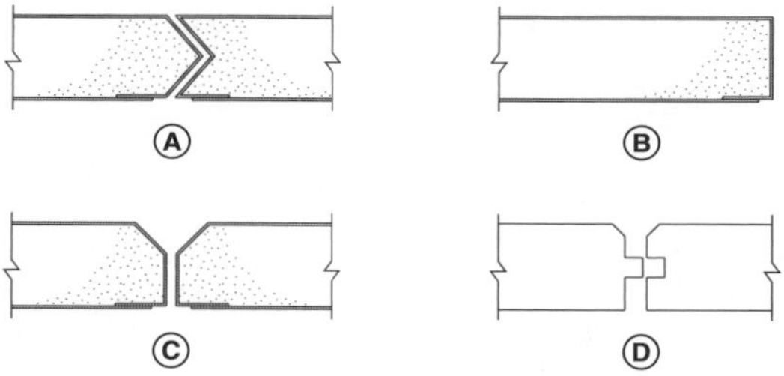

A B C D

Moisture-Resistant Drywall

_______________ **1.** Shower pan

_______________ **2.** ¼″ minimum air space

_______________ **3.** Moisture-resistant drywall

_______________ **4.** Nonhardening caulking

_______________ **5.** Ceramic tile

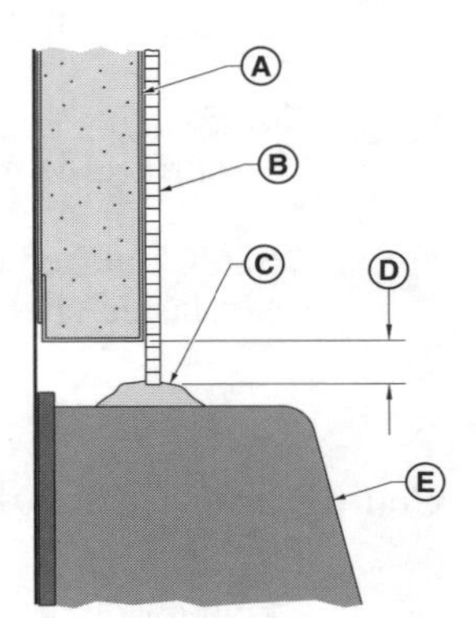

Gypsum Sheathing

_______________ **1.** Gypsum sheathing

_______________ **2.** Tex 1-11 siding

_______________ **3.** 15# asphalt-saturated felt

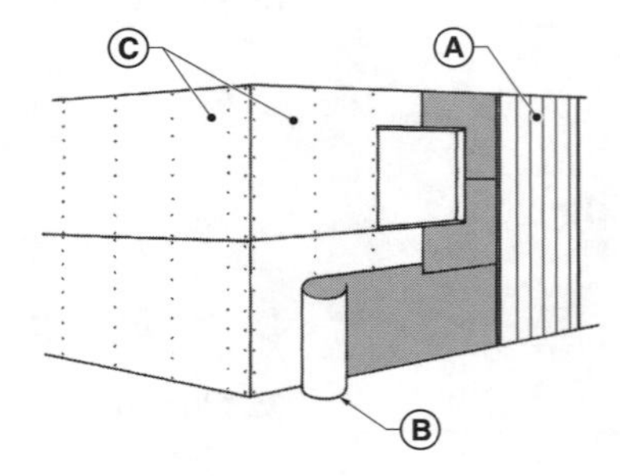

8 Taping, Finishing, and Texturing

Perforated paper tape and self-adhering fiberglass mesh tape are used for concealing the joints between drywall sheets. Perforated paper tape is used for high-production taping applications and is commonly applied with a Bazooka®. Two to three skim coats of joint topping compound are applied and sanded and nails are spotted to complete the drywall application.

TAPING

Taping is applying joint tape over joint compound to all drywall joints. Perforated paper tape and self-adhering fiberglass mesh tape are used for concealing the joints between drywall sheets. Tape may be applied with a Bazooka®, a banjo, or by hand.

Inspection

Properly-installed drywall sheets with all joints and fasteners concealed provide an ideal surface to receive a variety of textures, finishes, and wallcoverings. To ensure that all drywall joints and fasteners are properly concealed during the taping and finishing application, all areas to be finished must be inspected prior to installing the tape. The condition of the joints, faces, and fasteners of each drywall sheet must be carefully examined, and any condition that could adversely affect the finished product must be corrected. The drywall taper should be sure that:

1. All ends, edges, and faces of the drywall sheets are flat and free from fractures, bulges, or other damage.

2. The surface plane of adjoining drywall sheets is continuous except at the recess formed by two adjoining tapered edges or where it has been necessary to place one drywall sheet with a cut edge next to one with a tapered edge. See Figure 8-1.

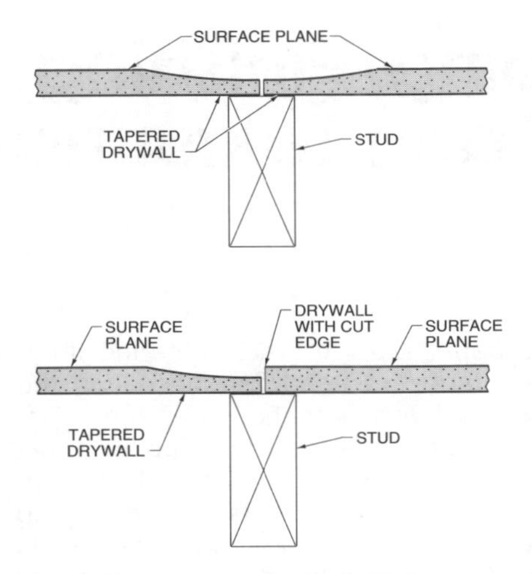

Figure 8-1. The surface plane of adjoining drywall sheets is continuous except at the recess formed by two adjoining sheets.

3. The space between the sides or ends of abutting drywall sheets does not exceed ¼″ in width.
4. All trim members have been properly installed and aligned.
5. All fasteners are installed properly and spaced correctly, and the heads of the fasteners are set slightly below the surface of the drywall. See Figure 8-2.

Most local building authorities require an inspection of the drywall fasteners before the taping operation can begin. When inspections are required, the drywall tapers must be sure that they have been completed and that the inspection card has been signed off before stringing tape or spotting nails.

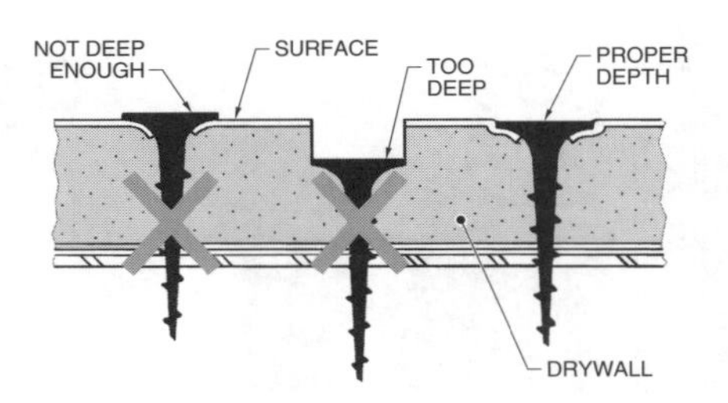

Figure 8-2. The heads of the fasteners are set slightly below the surface of the drywall.

Preparation

All fractured, dented, or otherwise damaged drywall edges, ends, or faces discovered during the inspection process must be cut out and the damaged drywall removed. The plane of all faces and adjoining drywall sheets must be continuous. Adjacent offset drywall sheets should be avoided. See Figure 8-3.

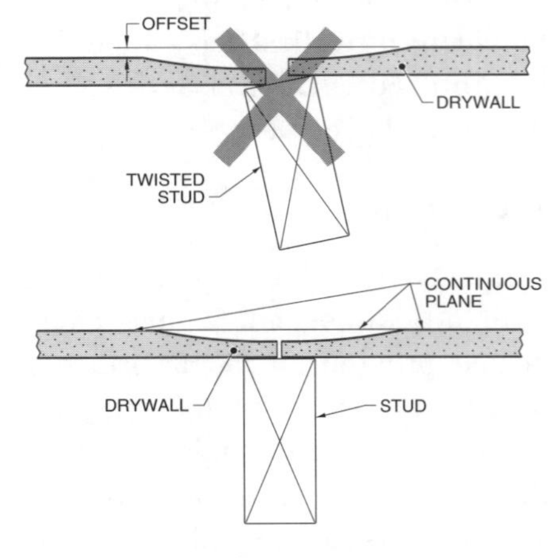

Figure 8-3. Adjacent offset drywall sheets should be avoided.

When severe bulges or offsets are discovered, it may be necessary to contact the drywall hangers and have the sheet removed to determine the cause of the problem. Whenever possible, all conditions requiring the removal of any drywall sheets should be corrected before the taping operation is begun. The need for additional taping and patching will be avoided if such corrections are made before taping begins.

All spaces between the drywall sheets which exceed 1/4″ and all voids created by the removal of damaged drywall must be filled with quick-setting joint compound before applying the drywall tape. See Figure 8-4. Drywall corner bead or other trim members not properly secured or aligned must be corrected by the drywall hanger. When the corner bead is installed by the taper, this step does not apply.

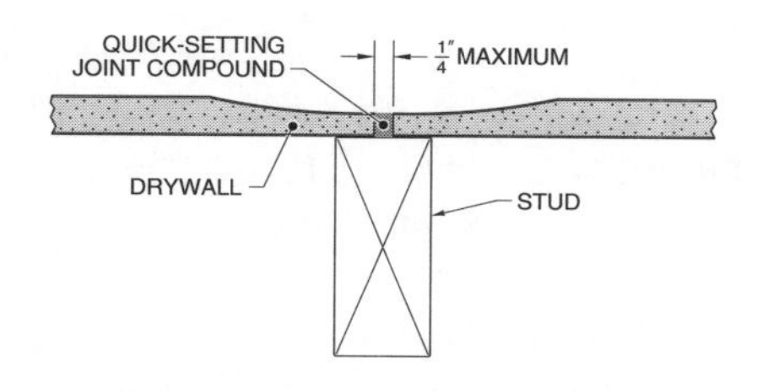

Figure 8-4. All spaces over 1/4″ must be filled with quick-setting joint compound before applying the drywall tape.

Screws or nails which are not set slightly below the surface must be corrected before beginning the nail-spotting process. Screws may be tightened with a Phillips screwdriver and nails may be driven in with the brass butt of the 6″ taping knife handle.

Installing Drywall Tape

Perforated paper tape and self-adhering fiberglass mesh tape are used for concealing the joints between drywall sheets. See Figure 8-5. Perforated paper tape is the most common drywall joint tape used for high-production taping applications. Self-adhering fiberglass mesh tape is ideal for low-production applications and drywall patching.

When installing perforated paper tape, apply a thin coating of joint compound (mud) to the surface of the drywall before the tape is installed. The joint compound may be applied with either of two mechanical taping machines. Both machines apply a thin coating of the joint compound and place the tape in position with one pass.

The Bazooka® is the most commonly used mechanical taping machine. It is used for high-production tape application. See Figure 8-6.

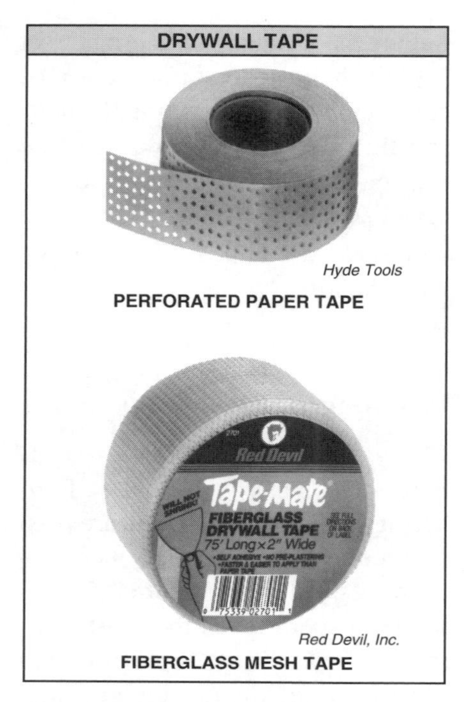

Figure 8-5. Perforated paper tape and self-adhering fiberglass mesh tape are used for concealing the joints between drywall sheets.

The banjo is used for low-production tape application and is ideal for patch work. See Figure 8-7. The final and least productive method for applying the thin coating of joint compound is spreading it by hand using a 6″ taping knife. Applying joint compound by hand requires the taper to spread a thin coating approximately 6″ wide on the joints of the drywall sheets for a distance of 8′ to 12′. Next, the paper tape is centered on the joint and pressed into the compound. Self-adhering fiberglass mesh tape is placed directly on the joints of the drywall sheets and does not require a thin coating of joint compound to hold it in place. With an adhesive coating on the back side, fiberglass mesh tape is ideal for small jobs and for patching drywall, plaster, and similar surfaces.

Figure 8-6. The Bazooka® is used for high-production tape application.

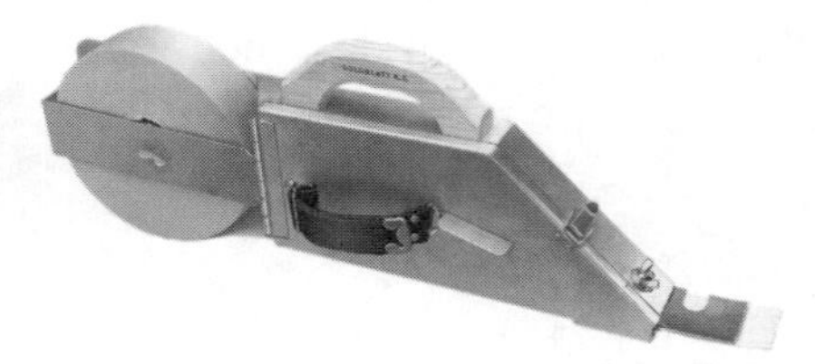

Stanley-Proto Industrial Tools

Figure 8-7. The banjo is used for low-production tape application.

When the Bazooka® will not be in continuous use for periods exceeding 15 min, it is advisable to place the head in a bucket of clean water to prevent the joint compound from drying out. At the end of the workday, the Bazooka® must be partially disassembled and thoroughly cleaned using a brush and clean water.

Wiping Down Tape. As the perforated paper tape is installed, it is necessary to press it into place centered on the joint and remove any excess joint compound. This process is referred to as wiping down the tape and is accomplished by using a standard 6″ or 8″ taping knife or a long handled wipe-down knife. See Figure 8-8.

National Gypsum Company

Figure 8-8. The tape is wiped down with a standard 6″ or 8″ taping knife.

Proper pressure ensures that the tape is properly seated and only the excess compound is removed. See Figure 8-9. If inadequate pressure is applied, excess joint compound remains beneath the tape and creates a bulge at the tape joint. If excessive pressure is applied, most of the joint compound is forced from beneath the tape and does not adhere to the surface of the drywall. When the wipe-down process is performed properly, the bridging of all drywall joints is achieved and the surface is prepared for succeeding coats of joint compound.

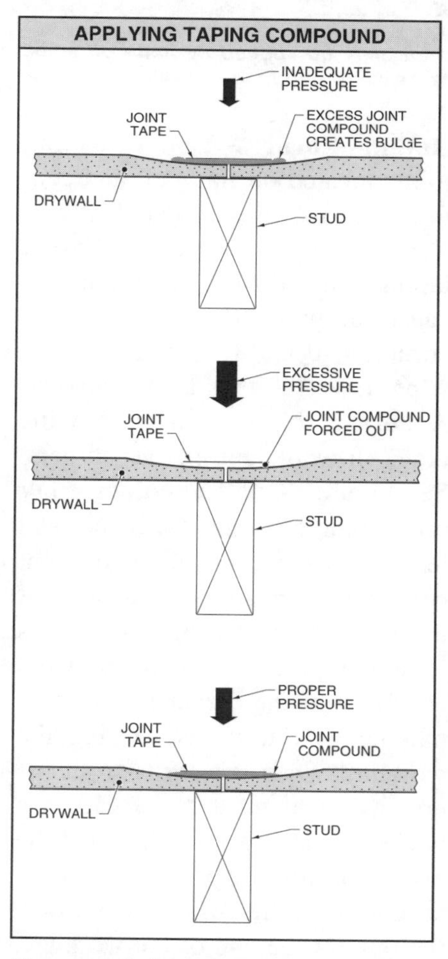

Figure 8-9. Proper pressure ensures that the tape is properly seated and only the excess joint compound is removed.

Skim coats may be applied by hand on small projects or when patching drywall.

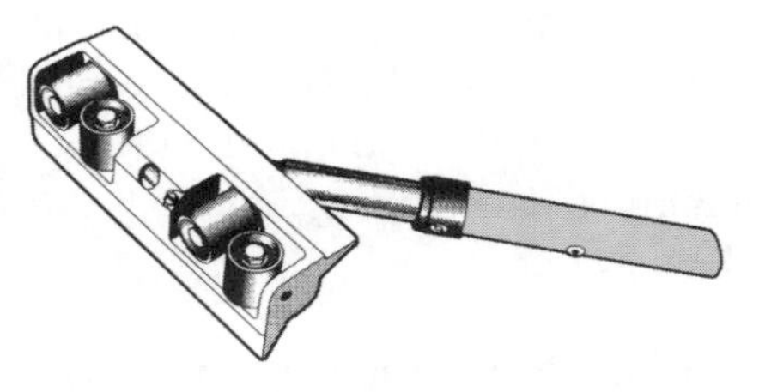

AMES Taping Tool Systems Inc.

Figure 8-10. A corner roller is used to wipe down joint tape in the corners.

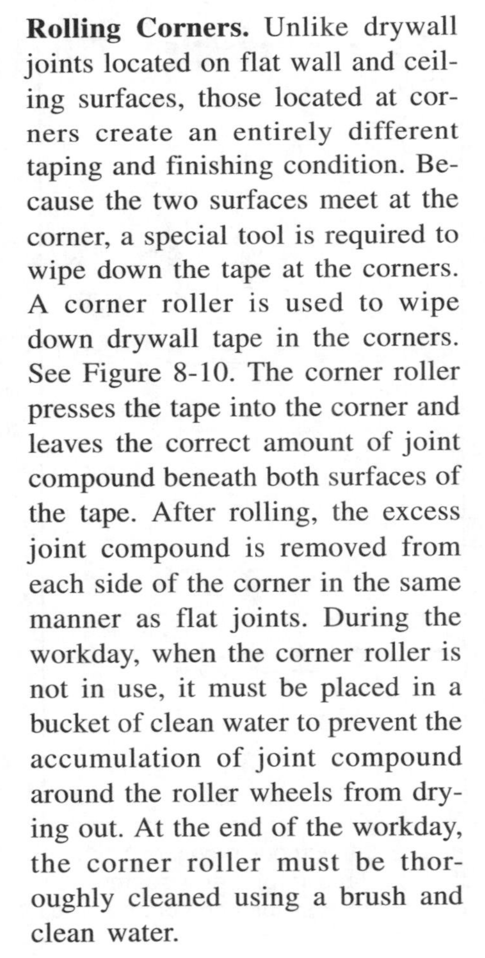

Rolling Corners. Unlike drywall joints located on flat wall and ceiling surfaces, those located at corners create an entirely different taping and finishing condition. Because the two surfaces meet at the corner, a special tool is required to wipe down the tape at the corners. A corner roller is used to wipe down drywall tape in the corners. See Figure 8-10. The corner roller presses the tape into the corner and leaves the correct amount of joint compound beneath both surfaces of the tape. After rolling, the excess joint compound is removed from each side of the corner in the same manner as flat joints. During the workday, when the corner roller is not in use, it must be placed in a bucket of clean water to prevent the accumulation of joint compound around the roller wheels from drying out. At the end of the workday, the corner roller must be thoroughly cleaned using a brush and clean water.

Fire Taping. *Fire taping* is drywall taping which is concealed behind paneling or above the finished ceiling to achieve the required fire rating. The joints must be taped and the nails spotted. Fire taping may be applied using either perforated paper tape or fiberglass mesh tape. When perforated paper tape is used for fire taping, a coating of joint compound may be required to completely cover the tape. The specific locations and requirements for fire taping are shown in the plans, listed in the specifications, or may be obtained from the AHJ. All fire taping requirements must be strictly followed. Fire tape may be applied using the Bazooka® or by hand, depending on the quantity and location.

Taping Exterior Ceiling and Soffit Drywall. Exterior ceiling and soffit drywall is taped using either perforated paper tape or fiberglass mesh tape and a special exterior grade joint compound. Exterior taping operations are similar to interior taping operations and may be performed using the Bazooka® and skim boxes or by hand.

Moisture-Resistant Drywall Taping. Moisture-resistant drywall, used as a backing sheet for tile in bathrooms and other high-moisture content areas, is taped using either perforated paper tape or fiberglass mesh tape and a special exterior grade joint compound. Perforated paper tape may be applied using the Bazooka® or by hand, depending on the quantity and location. If no tile is used and the moisture-resistant drywall must be finished for painting or wallcovering, the same taping and finishing techniques are employed as those used for regular drywall.

FINISHING

Finishing is begun after the joint compound is completely dry. Three skim coats are applied and sanded. A *skim coat* is a thin coat of joint topping compound. Skim coats can be applied by hand or with a mechanical skim box. See Figure 8-11. During the workday, when the skim boxes are not in use, they should be kept in a bucket of clean water to prevent the accumulation of joint topping compound from drying out. At the end of the workday, the skim boxes should be thoroughly cleaned using a brush and clean water.

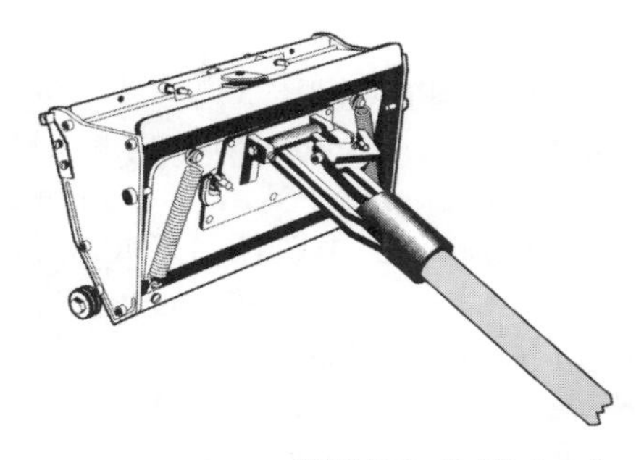

AMES Taping Tool Systems Inc.

Figure 8-11. A mechanical skim box is used to apply the joint topping compound for the skim coats.

Skim Coat No. 1

When the joint compound used to embed the perforated paper tape has thoroughly dried, the first skim coat of joint topping compound is applied. A 7″ mechanical skim box is used to apply the joint topping compound, or it may be applied by hand using the 8″ or 10″ taping knife. The mechanical skim box is ideal for high-production applications since it applies the desired thickness of joint topping compound to the joint surface and wipes it clean in one pass. Applying the first skim coat by hand with an 8″ or 10″ taping knife should be limited to small projects or when patching drywall.

Skim Coat No. 2

When the first skim coat of joint topping compound has thoroughly dried, a second skim coat of joint topping compound is applied. This second coat is applied using a 10″ or 12″ mechanical skim box or by hand using the 10″ or 12″ taping knife. The 10″ and 12″ mechanical skim boxes are ideal tools for high-production applications. They also apply the desired thickness of joint topping compound to the joint surface and remove any excess compound in one pass. The second skim coat may be applied with the 10″ or 12″ taping knife when working on small projects or when patching drywall.

Skim Coat No. 3

When a smooth finish is specified for the finished drywall surfaces, a third skim coat of joint topping compound is applied. This is done after the first and second skim coats have thoroughly dried. When a high rate of production is desired, the third skim coat is applied using a 12″ mechanical skim box. As with the 7″ and 10″ skim boxes, the 12″ skim box applies the desired thickness of joint topping compound and removes the excess in one pass. The third skim coat may also be applied with the 12″ taping knife on small projects or when patching drywall.

Nail Spotting

Nail spotting is covering the drywall screws or nails with several coats of joint topping compound. See Figure 8-12. Joint topping compound is applied to cover the fasteners while the skim coats are applied to the joints. Screws and nails are spotted using a mechanical nail spotting tool. Fasteners may also be spotted by hand using a 6″ taping knife. The mechanical nail spotting tool is used for high production. Hand nail spotting is used for small projects and drywall patching. Whichever method is employed, the object is to fill the dimples and to cover them with a thin coat of joint topping compound.

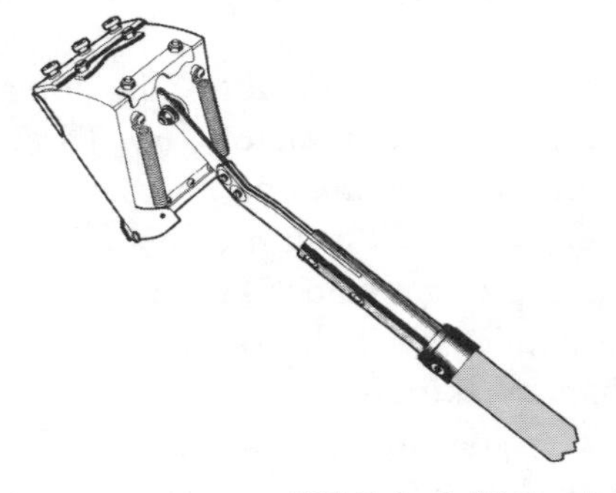

AMES Taping Tool Systems Inc.

Figure 8-12. Nail spotting is covering the drywall screws or nails with several coats of joint topping compound.

During the workday, when the nail spotter is not in use for extended periods of time, it should be kept in a bucket of clean water to prevent the joint topping compound from drying out. At the end of the workday, the nail spotter should be thoroughly cleaned using a brush and clean water.

Figure 8-13. Corner skimming is applying skim coats of joint topping compound to both surfaces of an inside corner joint.

Corner Skimming

Corner skimming is applying skim coats of joint topping compound to both surfaces of an inside corner joint. See Figure 8-13. The tape applied to cover the inside corner joints requires one or more skim coats of joint topping compound. The corner skimmer applies the correct amount of joint topping compound on both surfaces of the tape and removes the excess compound in one pass. Corner tape may also be coated by hand using a 6″ taping knife.

Like all other mechanical taping tools, when the corner skimmer is not in use, it should be placed in a bucket of clean water to prevent the joint compound from drying out. At the end of the workday, the corner skimmer should be thoroughly cleaned using a brush and clean water.

Coating Metal Trims

The perforated flanges of corner bead, L-metal, control joints, and other drywall trim are covered with two or three coats of joint compound. See Figure 8-14. All-purpose joint compound is used for the first coat. Joint topping compound is used for the succeeding one or two coats. The joint compound and joint topping compound is applied using a 6″ or 8″ taping knife. The bead of the trim member serves as a guide to determine the thickness of the compound being applied. Each coat is feathered out 6″ to 8″ from the bead of the trim member. Two coats of compound are required when texture is to be applied to the drywall surfaces and three coats are required when a smooth finish is desired.

Figure 8-14. The perforated flanges of metal trim are covered with two or three coats of joint compound.

Sanding

A sanding pole with a swivel head is used to sand the drywall joint compound between the first and second and the second and third skim coats. See Figure 8-15. The taped surfaces and the coating of joint compound on the metal trims may require some light sanding between applications, but the major sanding occurs after the final coats have been applied to the joints and the metal trims. Proper sanding ensures that the entire surface of the drywall is free from ridges or other blemishes. Care must be exercised when performing the sanding operation to avoid fuzzing the drywall face paper. This occurs when the sandpaper is allowed to drift off the edge of the mud coating and onto the drywall face paper. Excessive face paper fuzzing cannot be concealed with paint. Surfaces with excessive face paper fuzzing may require that a skim coat of joint topping compound be applied over the entire surface of the drywall.

Figure 8-15. A sanding pole with a swivel head is used to sand the joint compound between the first and second and the second and third skim coats.

Skim Coating Drywall Surfaces

Some drywall applications require that a skim coat of joint topping compound be applied over the entire surface of the drywall. This process adds considerable cost to the drywall application. However, when properly applied, skim coating a drywall surface provides a blemish-free surface suitable for painting or a variety of wallcoverings.

Skim coating of the entire surface of the drywall is applied after the second or third coat of joint compound has been applied to the joints, nails, and metal trims. A 12″ drywall knife is used to apply the skim coat. After the entire surface has thoroughly dried, it is given a final sanding. Any blemishes that remain are touched up and the surface is ready for decoration.

TEXTURING

Texturing is applying an uneven surface to the drywall. Walls and ceilings may be textured. A wide variety of wall textures are used to coat the entire drywall surface and prepare it for paint. These include spray textures of various patterns, skip-troweled texture, sand-finish textures using various grit sizes, combed textures, etc. See Figure 8-16. The building plans and specifications indicate the type of texture required.

The type of texture specified determines the degree to which the drywall joints, nails, and metal trims must be finished. Heavy textures require less attention when performing the basic taping and finishing operations. Light textures require nearly the same attention to detail as smooth wall finishes. When a variety of textures are being applied on the same project, it is advisable to mark the walls accordingly. This helps avoid unnecessary work being performed when it is not required. Likewise, when heavy wallcoverings are specified, less attention to detail is required than for thin wallcoverings. Many commercial drywall applications are finished with a light orange peel texture.

Spray Force Mfg., Inc.

Figure 8-16. Spray equipment is used to apply spray textures to drywall surfaces.

Residential Ceiling Texture

The most common ceiling texture applied in residential construction consists of a simulated acoustical spray material. This material, which requires no additional painting, is normally applied directly to the ceiling drywall surfaces in all rooms except the kitchen and bathrooms. Because simulated acoustical spray texture is considered a heavy texture, less attention to detail is required during the taping and finishing operations. Other textures are normally applied to the kitchen and bathroom ceilings, which receive paint.

Most residential drywall textures are applied using a spray machine that has two compartments. One compartment contains the ceiling texture material and the second compartment contains the wall texture material. These are often mounted on a large truck and are capable of applying texture to several complete residential units each day.

Commercial Ceiling Texture

Because most commercial drywall installations occur where the ceiling material specified is acoustical tile grid or other nondrywall material, the application of texture on commercial drywall ceilings is limited. The building plans and specifications indicate the areas that receive drywall ceilings and the finish that is to be applied to the surface of the drywall.

United States Gypsum Company

The Imperial™ QT Spray Texture Finish from United States Gypsum Company is an aggregated powder used on ceilings to conceal minor surface defects and produce an acoustical finish appearance.

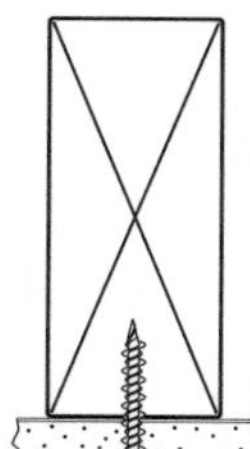

Taping, Finishing, and Texturing

Trade Competency Test

Name Date

Taping, Finishing, and Texturing

T F **1.** The space between the sides or ends of abutting drywall sheets should not exceed ¼″ in width.

__________ **2.** __________ is applying joint tape over joint compound to all drywall joints.

__________ **3.** Joint tape may be applied __________.

A. with a Bazooka®
B. by hand
C. with a banjo
D. A, B, and C

__________ **4.** __________ taping is drywall taping which is concealed behind paneling or above the finished ceiling to achieve the required fire rating.

T F **5.** All areas to be finished must be inspected prior to installing joint tape.

__________ **6.** All spaces between the drywall sheets which exceed ¼″ must be filled with __________ joint compound before applying the drywall tape.

__________ **7.** __________ tape is the most common drywall joint tape used for high-production taping applications.

__________ **8.** Screws or nails which are not set slightly below the surface must be corrected before beginning the __________ process.

T F **9.** Most local building authorities require an inspection of the drywall fasteners before the taping operation can begin.

_____________ **10.** When installing perforated paper tape, a thin coating of _________ is applied to the surface of the drywall before the tape is installed.

_____________ **11.** A(n) _________ roller is used to wipe down drywall tape in the corners.

_____________ **12.** _________ is the process of pressing joint tape on the joint to remove any excess joint compound.

_____________ **13.** Self adhering _________ mesh tape is ideal for low-production applications and drywall patching.

T F **14.** All fractured, dented, or otherwise damaged drywall edges, ends, or faces discovered during the inspection process must be completely covered with joint compound.

_____________ **15.** A(n) _________ is the most commonly used mechanical taping machine for high-production tape application.

_____________ **16.** A(n) _________ coat is a thin coat of joint topping compound.

_____________ **17.** _________ is applying skim coats of joint topping compound to both surfaces of an inside corner joint.

T F **18.** Finishing is begun before the joint compound is completely dry.

_____________ **19.** _________ is covering the drywall screws or nails with several coats of joint topping compound.

T F **20.** Skim coats may be applied by hand or with a mechanical skim box.

_____________ **21.** Drywall is finished by applying _________ skim coats of joint topping compound.

_____________ **22.** Corner tape may be coated by hand using a(n) _________″ taping knife.

_____________ **23.** A(n) _________ with a swivel head is used to sand the drywall joint compound between skim coats.

_____________ **24.** The third skim coat may be applied using a(n) _________″ taping knife on small projects or when patching drywall.

T F **25.** The first skim coat of joint topping compound is applied before the joint compound used to embed the perforated paper tape has dried.

___ **26.** ___ is applying an uneven surface to the drywall.

___ **27.** Many ___ drywall applications are finished with a light orange peel texture.

T F **28.** The most common ceiling texture applied in residential construction consists of a simulated acoustical spray material.

___ **29.** Spray textures used to coat entire drywall surfaces and prepare it for paint include ___ textures.

A. skip-troweled
B. combed
C. sand-finish
D. A, B, and C

T F **30.** More attention to detail is required during taping and finishing operations when drywall is to be covered with a heavy texture.

T F **31.** Whenever possible, all conditions requiring the removal of any drywall sheets should be corrected before the taping operation is begun.

___ **32.** Applying joint compound by hand requires the taper to spread a thin coating approximately 6″ wide on the joints of the drywall sheets for a distance of ___.

A. 4′ to 8′
B. 4′ to 12′
C. 8′ to 12′
D. neither A, B, nor C

___ **33.** When the Bazooka® will not be in continuous use for periods exceeding ___ min, it is advisable to place the head in a bucket of clean water to prevent the joint compound from drying out.

A. 5
B. 10
C. 15
D. 20

T F **34.** Fire tape may be applied using the Bazooka® or by hand, depending on the quantity and location.

T F **35.** The heads of drywall screws should be set at least ¼″ below the surface of the drywall.

Drywall Tools

_____________ **1.** A mechanical nail spotting tool is shown at _________.

_____________ **2.** A corner roller is shown at _________.

_____________ **3.** The tool shown at _________ is used to apply the joint topping compound for the skim coats.

_____________ **4.** A banjo is shown at _________.

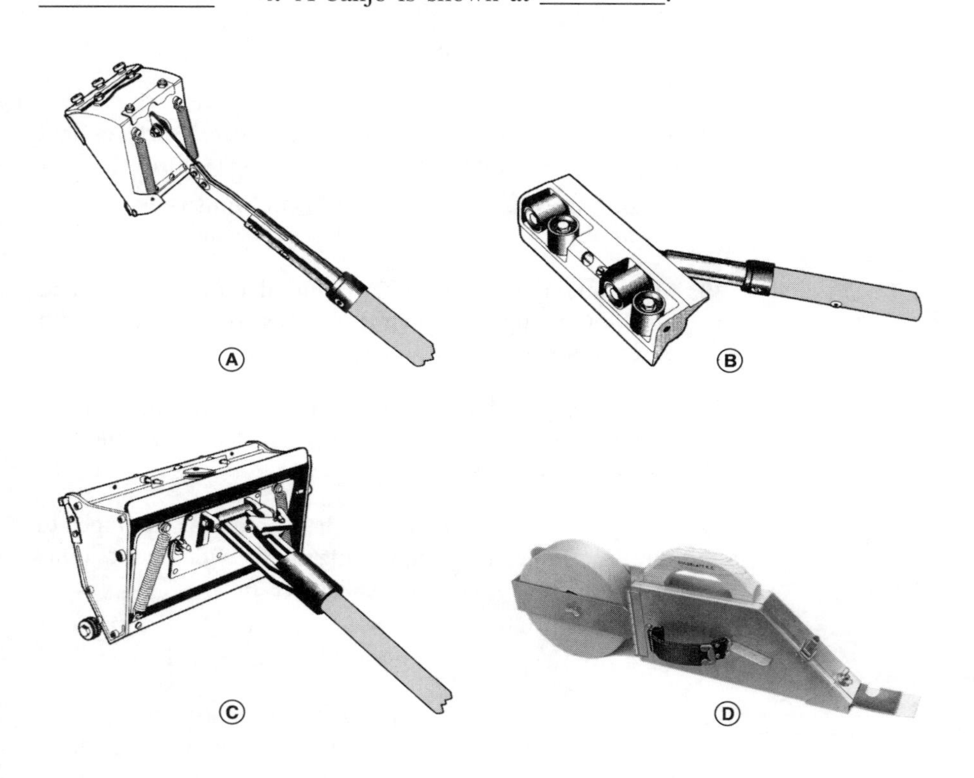

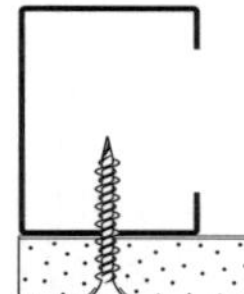

Appendix

Metal Trim Accessories 196
Common Gypsum Panel Sizes 196
Metal Track Sizes 197
Metal Stud Sizes 198
Decimal Equivalents of an Inch 198
Area – Plane Figures 199
Metrics 199
Metric System 200
English System 201
Architectural Symbols 202
Plot Plan Symbols 203
Electrical Symbols 204
Plumbing Symbols 206
HVAC Symbols 208

METAL TRIM ACCESSORIES	
Corner Bead Provides superior joint compound adhesion. Available in flange width: No. 103 1¼″ × 1¼″.	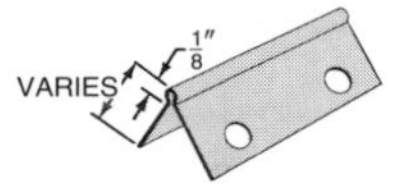
Expanded Flange Corner Bead Galvanized steel external corner reinforcement with 1/16″ grounds and 1¼″ wide fine-mesh expanded flanges. Provides superior key for joint compounds and eliminates shadowing and edge cracking.	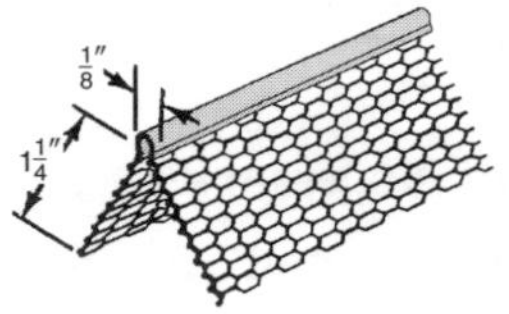
L-Trim and J-Trim Galvanized steel casing. J-shaped channel in ½″ and ⅝″ sizes. L-shaped angle edge trim without back flange to simplify application, in ½″ and ⅝″ sizes.	
J-Stop Reveal type galvanized steel trim. Requires no finishing compound, includes ⅜″, ½″, and ⅝″ size.	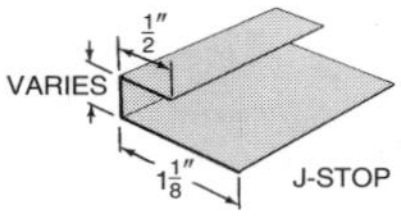
Expanded-Flange L-Trim and J-Trim Expanded-flange trim used to provide edge protection at cased openings and ceilings or wall intersections. J-shaped and L-shaped trim in ½″ and ⅝″ sizes.	

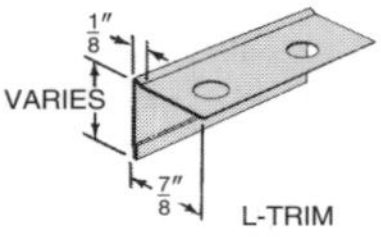

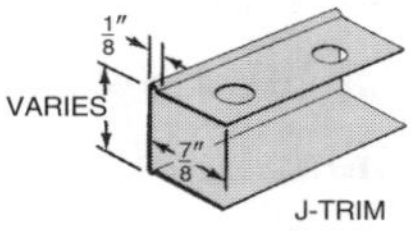

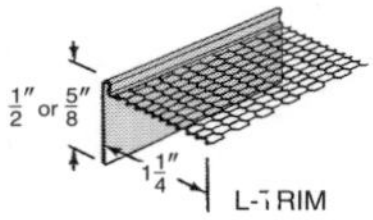

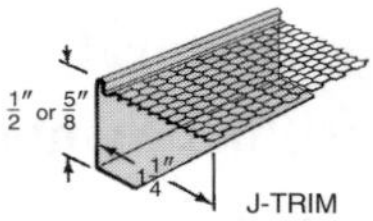

United States Gypsum Company

COMMON GYPSUM PANEL SIZES			
Gypsum Panel	**Thickness***	**Length****	**Width****
Regular	¼	8, 10	4
	⅜	8, 9, 10, 12, 14	4
	½, ⅝	8, 9, 10, 12, 14	4, 4½
Firecode	⅝	8, 9, 10, 12, 14	4
Foilback	½, ⅝	8, 9, 10, 12, 14	4
Water-Resistant	½, ⅝	8, 10, 12	4
Interior Ceiling	½	8, 12	4
Exterior Ceiling	½	8, 12	4
Vinyl Covered Panels	½, ⅝, ¾	8, 9, 10	2, 4
Gypsum Sheathing	½	8, 9	2, 4
	⅝	8, 9	4

* in in.

** in ft

METAL TRACK SIZES				
Depth*	**Gauge**	**Flanges***	**Weight/Ft****	**Length*****
1⅝	25	1	.24	10
		1¼	.27	
	22	1	.35	
		1¼	.4	
	20	1	.4	
		1¼	.46	
2½	25	1	.3	10
		1¼	.33	
	22	1	.44	
		1¼	.48	
	20	1	.5	
		1¼	.55	
3½	25	1	.36	10
		1¼	.39	
	22	1	.53	
		1¼	.58	
	20	1	.61	
		1¼	.66	
3⅝	25	1	.37	10
		1¼	.4	
	22	1	.54	
		1¼	.59	
	20	1	.62	
		1¼	.67	
4	25	1	.39	10
		1¼	.42	
	22	1	.58	
		1¼	.63	
	20	1	.66	
		1¼	.72	
6	25	1	.52	10
		1¼	.55	
	22	1	.77	
		1¼	.82	
	20	1	.88	
		1¼	.93	

* in in.

** in lb

*** in ft

METAL STUD SIZES					
Depth*	Flanges*	LIP*	Gauge	Weight/Ft**	Length***
1⅝	1 5/16, 1¼	.25	25	.29	20
			22	.43	
			20	.48	
2½	1 5/16, 1¼	.25	25	.35	20
			22	.51	
			20	.58	
3½	1 5/16, 1¼	.25	25	.41	20
			22	.61	
			20	.69	
3⅝	1 5/16, 1¼	.25	25	.42	20
			22	.62	
			20	.70	
4	1 5/16, 1¼	.25	25	.44	20
			22	.65	
			20	.74	
6	1 5/16, 1¼	.25	25	.57	20
			22	.84	
			20	.96	

* in in.
** in lb
*** in ft

DECIMAL EQUIVALENTS OF AN INCH							
Fraction	Decimal	Fraction	Decimal	Fraction	Decimal	Fraction	Decimal
1/64	0.015625	17/64	0.265625	33/64	0.515625	49/64	0.765625
1/32	0.03125	9/32	0.28125	17/32	0.53125	25/32	0.78125
3/64	0.046875	19/64	0.296875	35/64	0.546875	51/64	0.796875
1/16	0.0625	5/16	0.3125	9/16	0.5625	13/16	0.8125
5/64	0.078125	21/64	0.328125	37/64	0.578125	53/64	0.828125
3/32	0.09375	11/32	0.34375	19/32	0.59375	27/32	0.84375
7/64	0.109375	23/64	0.359375	39/64	0.609375	55/64	0.859375
1/8	0.125	3/8	0.375	5/8	0.625	7/8	0.875
9/64	0.140625	25/64	0.390625	41/64	0.640625	57/64	0.890625
5/32	0.15625	13/32	0.40625	21/32	0.65625	29/32	0.90625
11/64	0.171875	27/64	0.421875	43/64	0.671875	59/64	0.921875
3/16	0.1875	7/16	0.4375	11/16	0.6875	15/16	0.9375
13/64	0.203125	29/64	0.453125	45/64	0.703125	61/64	0.953125
7/32	0.21875	15/32	0.46875	23/32	0.71875	31/32	0.96875
15/64	0.234375	31/64	0.484375	47/64	0.734375	63/64	0.984375
1/4	0.250	1/2	0.500	3/4	0.750	1	1.000

AREA – PLANE FIGURES

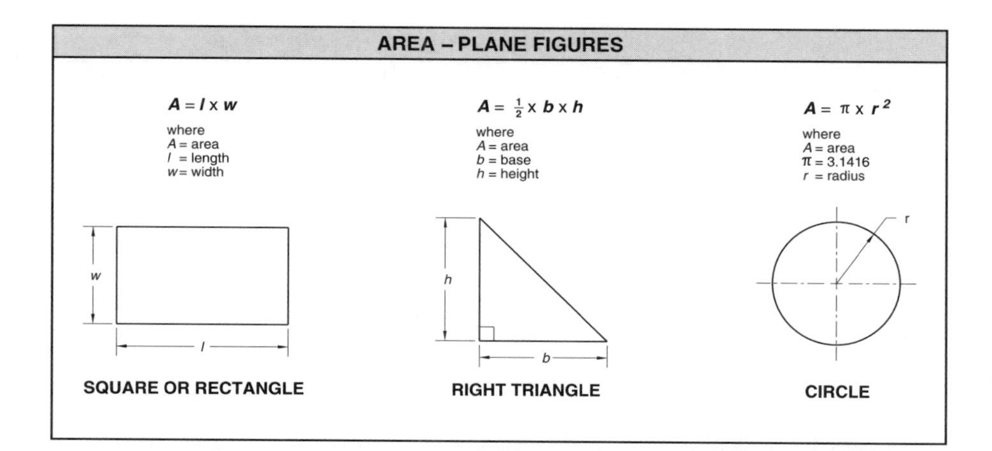

METRICS

BASE UNITS		
Unit	**SI Symbol**	**Quantity**
Meter	m	Length
Gram	g	Mass
Second	s	Time
Ampere	A	Electric current

UNIT PREFIXES			
Prefix	**Unit**	**Symbol**	**Number**
Other larger multiples			
Mega	Million	M	1,000,000 = 10^6
Kilo	Thousand	k	1,000 = 10^3
Hecto	Hundred	h	100 = 10^2
Deka	Ten	d	10 = 10^1
			Unit 1 = 10^0
Deci	Tenth	d	0.1 = 10^{-1}
Centi	Hundredth	c	0.01 = 10^{-2}
Milli	Thousandth	m	0.001 = 10^{-3}
Micro	Millionth	μ	0.000001 = 10^{-6}
Other smaller multiples			

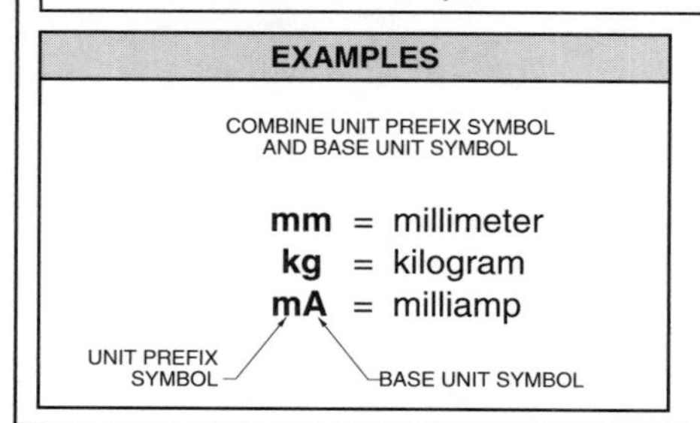

EXAMPLES
COMBINE UNIT PREFIX SYMBOL AND BASE UNIT SYMBOL
mm = millimeter
kg = kilogram
mA = milliamp
UNIT PREFIX SYMBOL
BASE UNIT SYMBOL

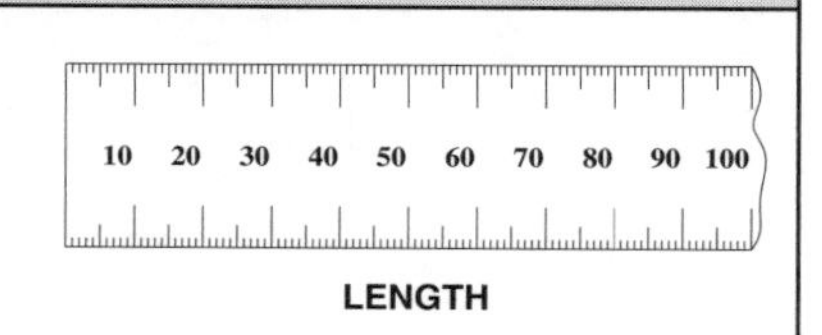

LENGTH

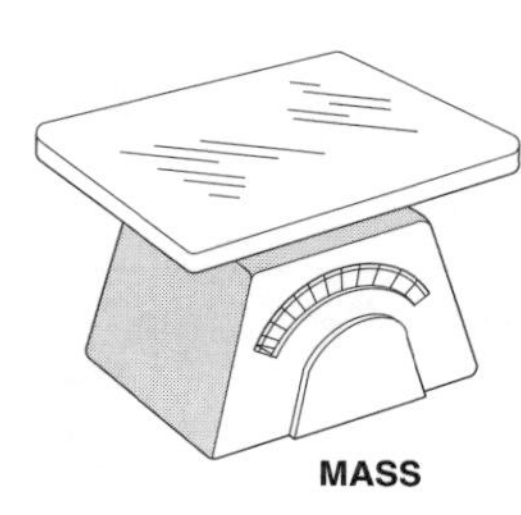

MASS

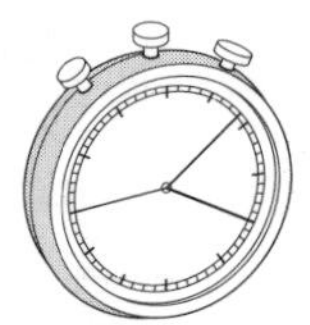

TIME

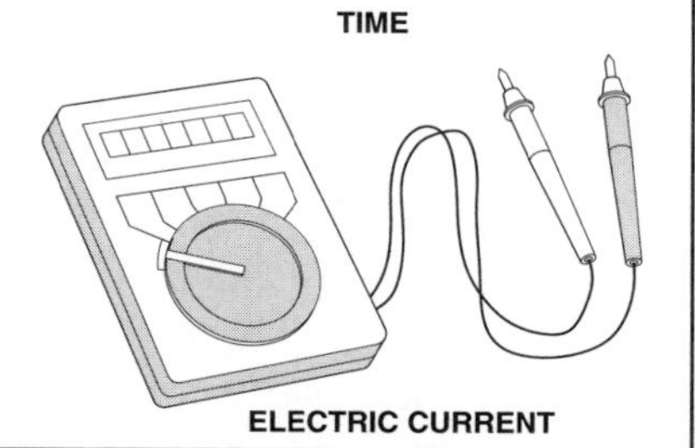

ELECTRIC CURRENT

METRIC SYSTEM			
LENGTH	**Unit**	**Abbr**	**Number of Base Units**
	kilometer	km	1000
	hectometer	hm	100
	dekameter	dam	10
	***meter**	m	1
	decimeter	dm	.1
	centimeter	cm	.01
	millimeter	mm	.001
AREA	square kilometer	sq km *or* km^2	1,000,000
	hectare	ha	10,000
	are	a	100
	square centimeter	sq cm *or* cm^2	.0001
VOLUME	cubic centimeter	cu cm, cm^3, *or* cc	.000001
	cubic decimeter	dm^3	.001
	***cubic meter**	m^3	1
CAPACITY	kiloliter	kl	1000
	hectoliter	hl	100
	dekaliter	dal	10
	***liter**	l	1
	cubic decimeter	dm^3	1
	deciliter	dl	.10
	centiliter	cl	.01
	milliliter	ml	.001
MASS AND WEIGHT	metric ton	t	1,000,000
	kilogram	kg	1000
	hectogram	hg	100
	dekagram	dag	10
	***gram**	g	1
	decigram	dg	.10
	centigram	cg	.01
	milligram	mg	.001

* base units

ENGLISH SYSTEM				
LENGTH		**Unit**	**Abbr**	**Equivalents**
		mile	mi	5280′, 320 rd, 1760 yd
		rod	rd	5.50 yd, 16.5′
		yard	yd	3′, 36″
		foot	ft *or* ′	12″, .333 yd
		inch	in. *or* ″	.083′, .028 yd
AREA A = l x w		square mile	sq mi *or* mi^2	640 A, 102,400 sq rd
		acre	A	4840 sq yd, 43,560 sq ft
		square rod	sq rd *or* rd^2	30.25 sq yd, .00625 A
		square yard	sq yd *or* yd^2	1296 sq in., 9 sq ft
		square foot	sq ft *or* ft^2	144 sq in., .111 sq yd
		square inch	sq in. *or* in^2	.0069 sq ft, .00077 sq yd
VOLUME V = l x w x t		cubic yard	cu yd *or* yd^3	27 cu ft, 46,656 cu in.
		cubic foot	cu ft *or* ft^3	1728 cu in., .0370 cu yd
		cubic inch	cu in. *or* in^3	.00058 cu ft, .000021 cu yd
CAPACITY WATER, FUEL, ETC.	*U.S. liquid measure*	gallon	gal.	4 qt (231 cu in.)
		quart	qt	2 pt (57.75 cu in.)
		pint	pt	4 gi (28.875 cu in.)
		gill	gi	4 fl oz (7.219 cu in.)
		fluidounce	fl oz	8 fl dr (1.805 cu in.)
		fluidram	fl dr	60 min (.226 cu in.)
		minim	min	1⁄6 fl dr (.003760 cu in.)
VEGETABLES, GRAIN, ETC.	*U.S. dry measure*	bushel	bu	4 pk (2150.42 cu in.)
		peck	pk	8 qt (537.605 cu in.)
		quart	qt	2 pt (67.201 cu in.)
		pint	pt	1⁄2 qt (33.600 cu in.)
DRUGS	*British imperial liquid and dry measure*	bushel	bu	4 pk (2219.36 cu in.)
		peck	pk	2 gal. (554.84 cu in.)
		gallon	gal.	4 qt (277.420 cu in.)
		quart	qt	2 pt (69.355 cu in.)
		pint	pt	4 gi (34.678 cu in.)
		gill	gi	5 fl oz (8.669 cu in.)
		fluidounce	fl oz	8 fl dr (1.7339 cu in.)
		fluidram	fl dr	60 min (.216734 cu in.)
		minim	min	1⁄60 fl dr (.003612 cu in.)
MASS AND WEIGHT COAL, GRAIN, ETC.	*avoirdupois*	ton	t	2000 lb
		short ton		2000 lb
		long ton		2240 lb
		pound	lb *or* #	16 oz, 7000 gr
		ounce	oz	16 dr, 437.5 gr
		dram	dr	27.344 gr, .0625 oz
		grain	gr	.037 dr, .002286 oz
GOLD, SILVER, ETC.	*troy*	pound	lb	12 oz, 240 dwt, 5760 gr
		ounce	oz	20 dwt, 480 gr
		pennyweight	dwt *or* pwt	24 gr, .05 oz
		grain	gr	.042 dwt, .002083 oz
DRUGS	*apothecaries'*	pound	lb ap	12 oz, 5760 gr
		ounce	oz ap	8 dr ap, 480 gr
		dram	dr ap	3 s ap, 60 gr
		scruple	s ap	20 gr, .333 dr ap
		grain	gr	.05 s, .002083 oz, .0166 dr ap

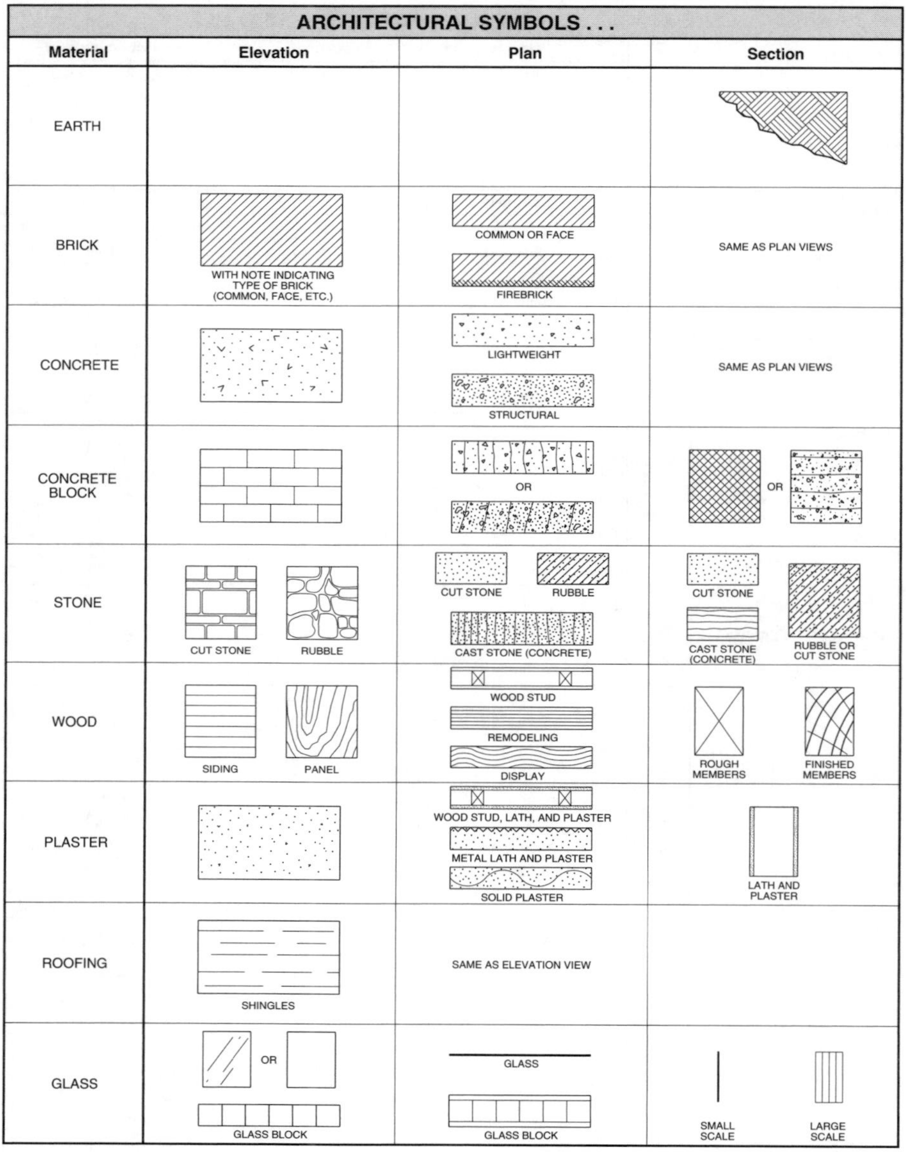

ARCHITECTURAL SYMBOLS . . .			
Material	**Elevation**	**Plan**	**Section**
EARTH			
BRICK	WITH NOTE INDICATING TYPE OF BRICK (COMMON, FACE, ETC.)	COMMON OR FACE FIREBRICK	SAME AS PLAN VIEWS
CONCRETE		LIGHTWEIGHT STRUCTURAL	SAME AS PLAN VIEWS
CONCRETE BLOCK		OR	OR
STONE	CUT STONE RUBBLE	CUT STONE RUBBLE CAST STONE (CONCRETE)	CUT STONE CAST STONE (CONCRETE) RUBBLE OR CUT STONE
WOOD	SIDING PANEL	WOOD STUD REMODELING DISPLAY	ROUGH MEMBERS FINISHED MEMBERS
PLASTER		WOOD STUD, LATH, AND PLASTER METAL LATH AND PLASTER SOLID PLASTER	LATH AND PLASTER
ROOFING	SHINGLES	SAME AS ELEVATION VIEW	
GLASS	OR GLASS BLOCK	GLASS GLASS BLOCK	SMALL SCALE LARGE SCALE

. . . ARCHITECTURAL SYMBOLS

Material	Elevation	Plan	Section
FACING TILE	CERAMIC TILE	FLOOR TILE	CERAMIC TILE LARGE SCALE; CERAMIC TILE SMALL SCALE
STRUCTURAL CLAY TILE			SAME AS PLAN VIEW
INSULATION		LOOSE FILL OR BATTS; RIGID; SPRAY FOAM	SAME AS PLAN VIEWS
SHEET METAL FLASHING		OCCASIONALLY INDICATED BY NOTE	
METALS OTHER THAN FLASHING	INDICATED BY NOTE OR DRAWN TO SCALE	SAME AS ELEVATION	SMALL SCALE; STEEL; CAST IRON; ALUMINUM; BRONZE OR BRASS
STRUCTURAL STEEL	INDICATED BY NOTE OR DRAWN TO SCALE	OR	REBARS; SMALL SCALE; LARGE SCALE; L-ANGLES, S-BEAMS, ETC.

PLOT PLAN SYMBOLS

N	NORTH		FIRE HYDRANT		WALK	E OR	ELECTRIC SERVICE
	POINT OF BEGINNING (POB)		MAILBOX		IMPROVED ROAD	G OR	NATURAL GAS LINE
	UTILITY METER OR VALVE		MANHOLE		UNIMPROVED ROAD	W OR	WATER LINE
	POWER POLE AND GUY		TREE		BUILDING LINE	T OR	TELEPHONE LINE
	LIGHT STANDARD		BUSH		PROPERTY LINE		NATURAL GRADE
	TRAFFIC SIGNAL		HEDGE ROW		PROPERTY LINE		FINISH GRADE
	STREET SIGN		FENCE		TOWNSHIP LINE	+ XX.00′	EXISTING ELEVATION

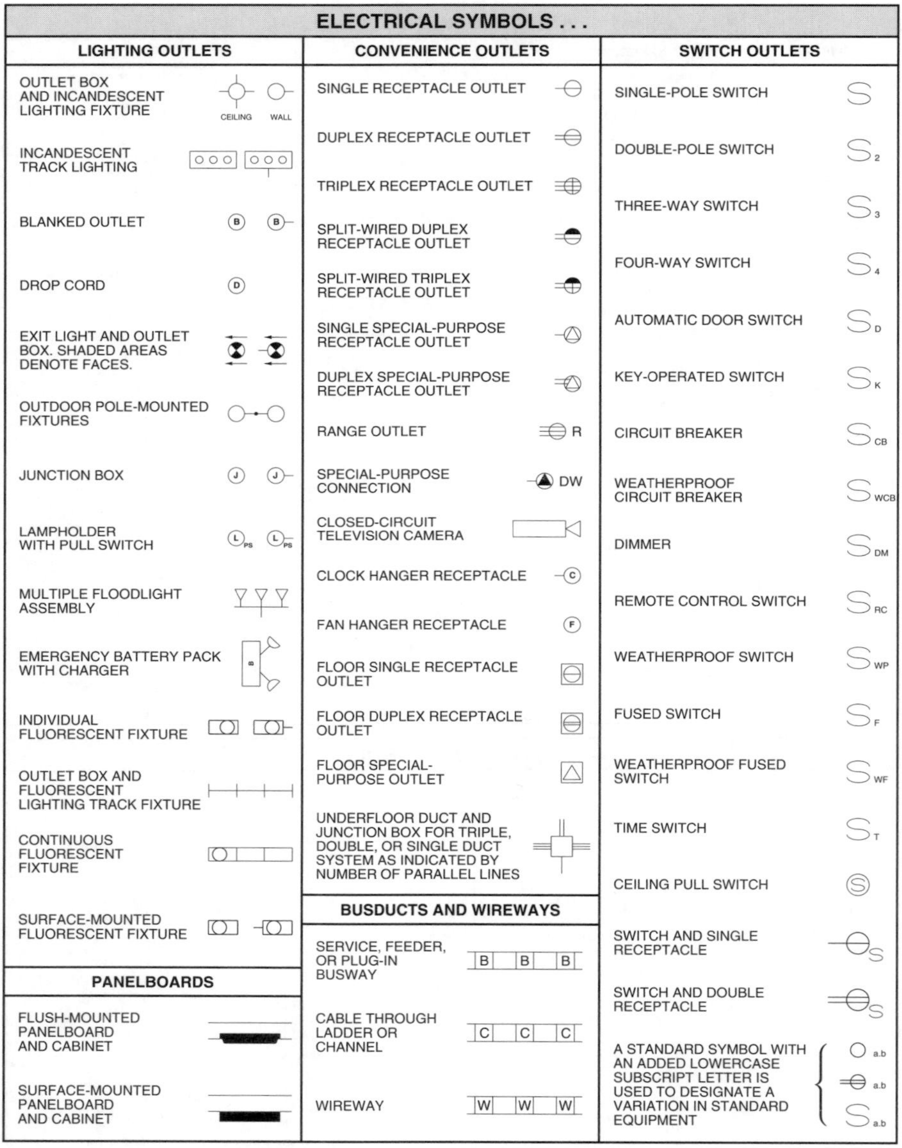

ELECTRICAL SYMBOLS . . .

LIGHTING OUTLETS

OUTLET BOX AND INCANDESCENT LIGHTING FIXTURE	CEILING / WALL
INCANDESCENT TRACK LIGHTING	
BLANKED OUTLET	B / B
DROP CORD	D
EXIT LIGHT AND OUTLET BOX. SHADED AREAS DENOTE FACES.	
OUTDOOR POLE-MOUNTED FIXTURES	
JUNCTION BOX	J / J
LAMPHOLDER WITH PULL SWITCH	L_{PS} / L_{PS}
MULTIPLE FLOODLIGHT ASSEMBLY	
EMERGENCY BATTERY PACK WITH CHARGER	B
INDIVIDUAL FLUORESCENT FIXTURE	
OUTLET BOX AND FLUORESCENT LIGHTING TRACK FIXTURE	
CONTINUOUS FLUORESCENT FIXTURE	
SURFACE-MOUNTED FLUORESCENT FIXTURE	

PANELBOARDS

FLUSH-MOUNTED PANELBOARD AND CABINET	
SURFACE-MOUNTED PANELBOARD AND CABINET	

CONVENIENCE OUTLETS

SINGLE RECEPTACLE OUTLET	
DUPLEX RECEPTACLE OUTLET	
TRIPLEX RECEPTACLE OUTLET	
SPLIT-WIRED DUPLEX RECEPTACLE OUTLET	
SPLIT-WIRED TRIPLEX RECEPTACLE OUTLET	
SINGLE SPECIAL-PURPOSE RECEPTACLE OUTLET	
DUPLEX SPECIAL-PURPOSE RECEPTACLE OUTLET	
RANGE OUTLET	R
SPECIAL-PURPOSE CONNECTION	DW
CLOSED-CIRCUIT TELEVISION CAMERA	
CLOCK HANGER RECEPTACLE	C
FAN HANGER RECEPTACLE	F
FLOOR SINGLE RECEPTACLE OUTLET	
FLOOR DUPLEX RECEPTACLE OUTLET	
FLOOR SPECIAL-PURPOSE OUTLET	
UNDERFLOOR DUCT AND JUNCTION BOX FOR TRIPLE, DOUBLE, OR SINGLE DUCT SYSTEM AS INDICATED BY NUMBER OF PARALLEL LINES	

BUSDUCTS AND WIREWAYS

SERVICE, FEEDER, OR PLUG-IN BUSWAY	B B B
CABLE THROUGH LADDER OR CHANNEL	C C C
WIREWAY	W W W

SWITCH OUTLETS

SINGLE-POLE SWITCH	S
DOUBLE-POLE SWITCH	S_2
THREE-WAY SWITCH	S_3
FOUR-WAY SWITCH	S_4
AUTOMATIC DOOR SWITCH	S_D
KEY-OPERATED SWITCH	S_K
CIRCUIT BREAKER	S_{CB}
WEATHERPROOF CIRCUIT BREAKER	S_{WCB}
DIMMER	S_{DM}
REMOTE CONTROL SWITCH	S_{RC}
WEATHERPROOF SWITCH	S_{WP}
FUSED SWITCH	S_F
WEATHERPROOF FUSED SWITCH	S_{WF}
TIME SWITCH	S_T
CEILING PULL SWITCH	S
SWITCH AND SINGLE RECEPTACLE	S
SWITCH AND DOUBLE RECEPTACLE	S
A STANDARD SYMBOL WITH AN ADDED LOWERCASE SUBSCRIPT LETTER IS USED TO DESIGNATE A VARIATION IN STANDARD EQUIPMENT	a.b / a.b / $S_{a.b}$

. . . ELECTRICAL SYMBOLS

COMMERCIAL AND INDUSTRIAL SYSTEMS

PAGING SYSTEM DEVICE

FIRE ALARM SYSTEM DEVICE

COMPUTER DATA SYSTEM DEVICE

PRIVATE TELEPHONE SYSTEM DEVICE

SOUND SYSTEM

FIRE ALARM CONTROL PANEL — FACP

SIGNALING SYSTEM OUTLETS FOR RESIDENTIAL SYSTEMS

PUSHBUTTON

BUZZER

BELL

BELL AND BUZZER COMBINATION

COMPUTER DATA OUTLET

BELL RINGING TRANSFORMER — BT

ELECTRIC DOOR OPENER — D

CHIME — CH

TELEVISION OUTLET — TV

THERMOSTAT — T

UNDERGROUND ELECTRICAL DISTRIBUTION OR ELECTRICAL LIGHTING SYSTEMS

MANHOLE — M

HANDHOLE — H

TRANSFORMER-MANHOLE OR VAULT — TM

TRANSFORMER PAD — TP

UNDERGROUND DIRECT BURIAL CABLE

UNDERGROUND DUCT LINE

STREET LIGHT STANDARD FED FROM UNDERGROUND CIRCUIT

ABOVE-GROUND ELECTRICAL DISTRIBUTION OR LIGHTING SYSTEMS

POLE

STREET LIGHT AND BRACKET

PRIMARY CIRCUIT

SECONDARY CIRCUIT

DOWN GUY

HEAD GUY

SIDEWALK GUY

SERVICE WEATHERHEAD

PANEL CIRCUITS AND MISCELLANEOUS

LIGHTING PANEL

POWER PANEL

WIRING – CONCEALED IN CEILING OR WALL

WIRING – CONCEALED IN FLOOR

WIRING EXPOSED

HOME RUN TO PANEL BOARD
Indicate number of circuits by number of arrows. Any circuit without such designation indicates a two-wire circuit. For a greater number of wires indicate as follows: (3 wires)
(4 wires), etc.

FEEDERS
Use heavy lines and designate by number corresponding to listing in feeder schedule

WIRING TURNED UP

WIRING TURNED DOWN

GENERATOR — G

MOTOR — M

INSTRUMENT (SPECIFY) — I

TRANSFORMER — T

CONTROLLER

EXTERNALLY-OPERATED DISCONNECT SWITCH

PULL BOX

PLUMBING SYMBOLS . . .

FIXTURES...	
STANDARD BATHTUB	
OVAL BATHTUB	
WHIRLPOOL BATH	
SHOWER STALL	
SHOWER HEAD	
TANK-TYPE WATER CLOSET	
WALL-MOUNTED WATER CLOSET	
FLOOR-MOUNTED WATER CLOSET	
LOW-PROFILE WATER CLOSET	
BIDET	
WALL-MOUNTED URINAL	
FLOOR-MOUNTED URINAL	
TROUGH-TYPE URINAL	
WALL-MOUNTED LAVATORY	
PEDESTAL LAVATORY	
BUILT-IN LAVATORY	
WHEELCHAIR LAVATORY	
CORNER LAVATORY	
FLOOR DRAIN	
FLOOR SINK	

...FIXTURES	
LAUNDRY TRAY	
BUILT-IN SINK	
DOUBLE OR TRIPLE BUILT-IN SINK	
COMMERCIAL KITCHEN SINK	
SERVICE SINK	SS
CLINIC SERVICE SINK	
FLOOR-MOUNTED SERVICE SINK	
DRINKING FOUNTAIN	DF
WATER COOLER	
HOT WATER TANK	HW T
WATER HEATER	WH
METER	M
HOSE BIBB	HB
GAS OUTLET	G
GREASE SEPARATOR	G
GARAGE DRAIN	
FLOOR DRAIN WITH BACKWATER VALVE	

PIPING...	
SOIL, WASTE, OR LEADER – ABOVE GRADE	
SOIL, WASTE, OR LEADER – BELOW GRADE	
VENT	
COMBINATION WASTE AND VENT	SV
STORM DRAIN	SD
COLD WATER	

...PIPING	
CHILLED DRINKING WATER SUPPLY	DWS
CHILLED DRINKING WATER RETURN	DWR
HOT WATER	
HOT WATER RETURN	
SANITIZING HOT WATER SUPPLY (180°F)	
SANITIZING HOT WATER RETURN (180°F)	
DRY STANDPIPE	DSP
COMBINATION STANDPIPE	CSP
MAIN SUPPLIES SPRINKLER	S
BRANCH AND HEAD SPRINKLER	
GAS – LOW PRESSURE	G G
GAS – MEDIUM PRESSURE	MG
GAS – HIGH PRESSURE	HG
COMPRESSED AIR	A
OXYGEN	O
NITROGEN	N
HYDROGEN	H
HELIUM	HE
ARGON	AR
LIQUID PETROLEUM GAS	LPG
INDUSTRIAL WASTE	INW
CAST IRON	CI
CULVERT PIPE	CP
CLAY TILE	CT
DUCTILE IRON	DI
REINFORCED CONCRETE	RCP
DRAIN – OPEN TILE OR AGRICULTURAL TILE	

. . . PLUMBING SYMBOLS

PIPE FITTING AND VALVE SYMBOLS

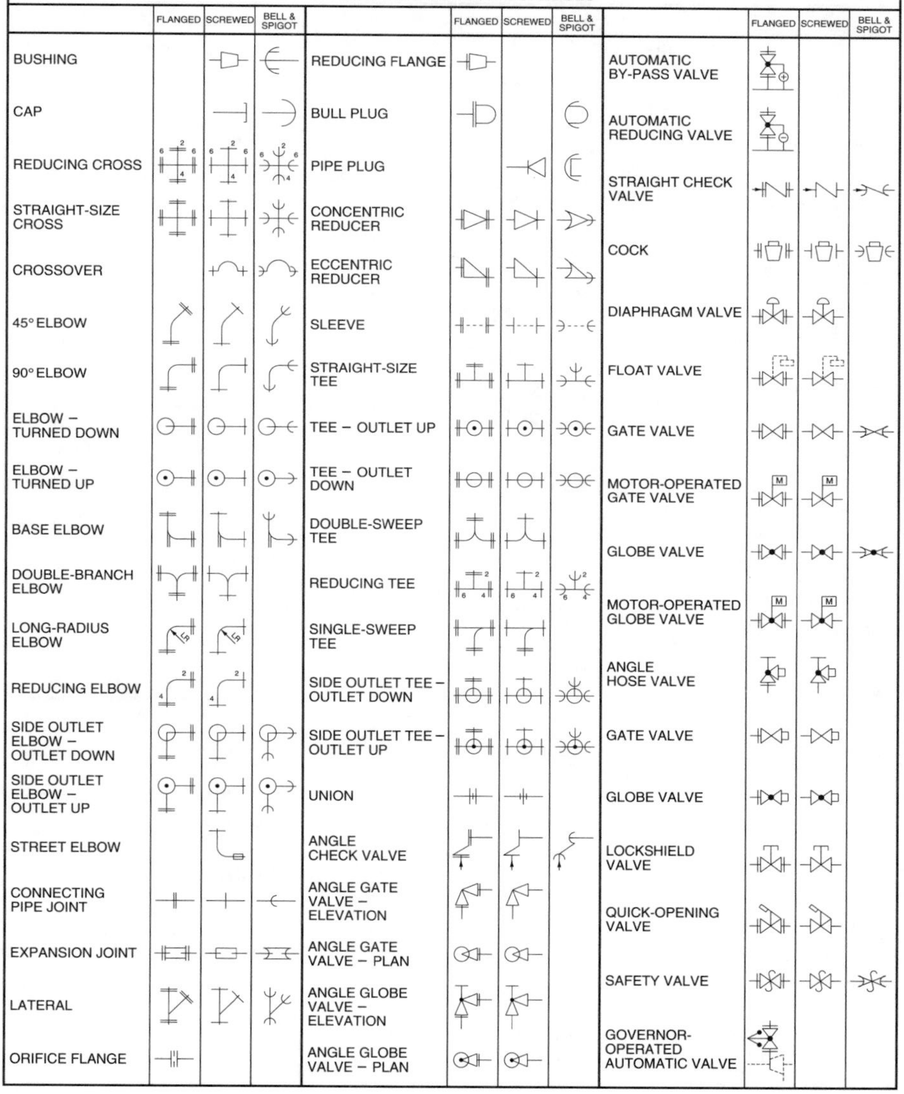

	FLANGED	SCREWED	BELL & SPIGOT
BUSHING			
CAP			
REDUCING CROSS			
STRAIGHT-SIZE CROSS			
CROSSOVER			
45° ELBOW			
90° ELBOW			
ELBOW – TURNED DOWN			
ELBOW – TURNED UP			
BASE ELBOW			
DOUBLE-BRANCH ELBOW			
LONG-RADIUS ELBOW			
REDUCING ELBOW			
SIDE OUTLET ELBOW – OUTLET DOWN			
SIDE OUTLET ELBOW – OUTLET UP			
STREET ELBOW			
CONNECTING PIPE JOINT			
EXPANSION JOINT			
LATERAL			
ORIFICE FLANGE			

	FLANGED	SCREWED	BELL & SPIGOT
REDUCING FLANGE			
BULL PLUG			
PIPE PLUG			
CONCENTRIC REDUCER			
ECCENTRIC REDUCER			
SLEEVE			
STRAIGHT-SIZE TEE			
TEE – OUTLET UP			
TEE – OUTLET DOWN			
DOUBLE-SWEEP TEE			
REDUCING TEE			
SINGLE-SWEEP TEE			
SIDE OUTLET TEE – OUTLET DOWN			
SIDE OUTLET TEE – OUTLET UP			
UNION			
ANGLE CHECK VALVE			
ANGLE GATE VALVE – ELEVATION			
ANGLE GATE VALVE – PLAN			
ANGLE GLOBE VALVE – ELEVATION			
ANGLE GLOBE VALVE – PLAN			

	FLANGED	SCREWED	BELL & SPIGOT
AUTOMATIC BY-PASS VALVE			
AUTOMATIC REDUCING VALVE			
STRAIGHT CHECK VALVE			
COCK			
DIAPHRAGM VALVE			
FLOAT VALVE			
GATE VALVE			
MOTOR-OPERATED GATE VALVE			
GLOBE VALVE			
MOTOR-OPERATED GLOBE VALVE			
ANGLE HOSE VALVE			
GATE VALVE			
GLOBE VALVE			
LOCKSHIELD VALVE			
QUICK-OPENING VALVE			
SAFETY VALVE			
GOVERNOR-OPERATED AUTOMATIC VALVE			

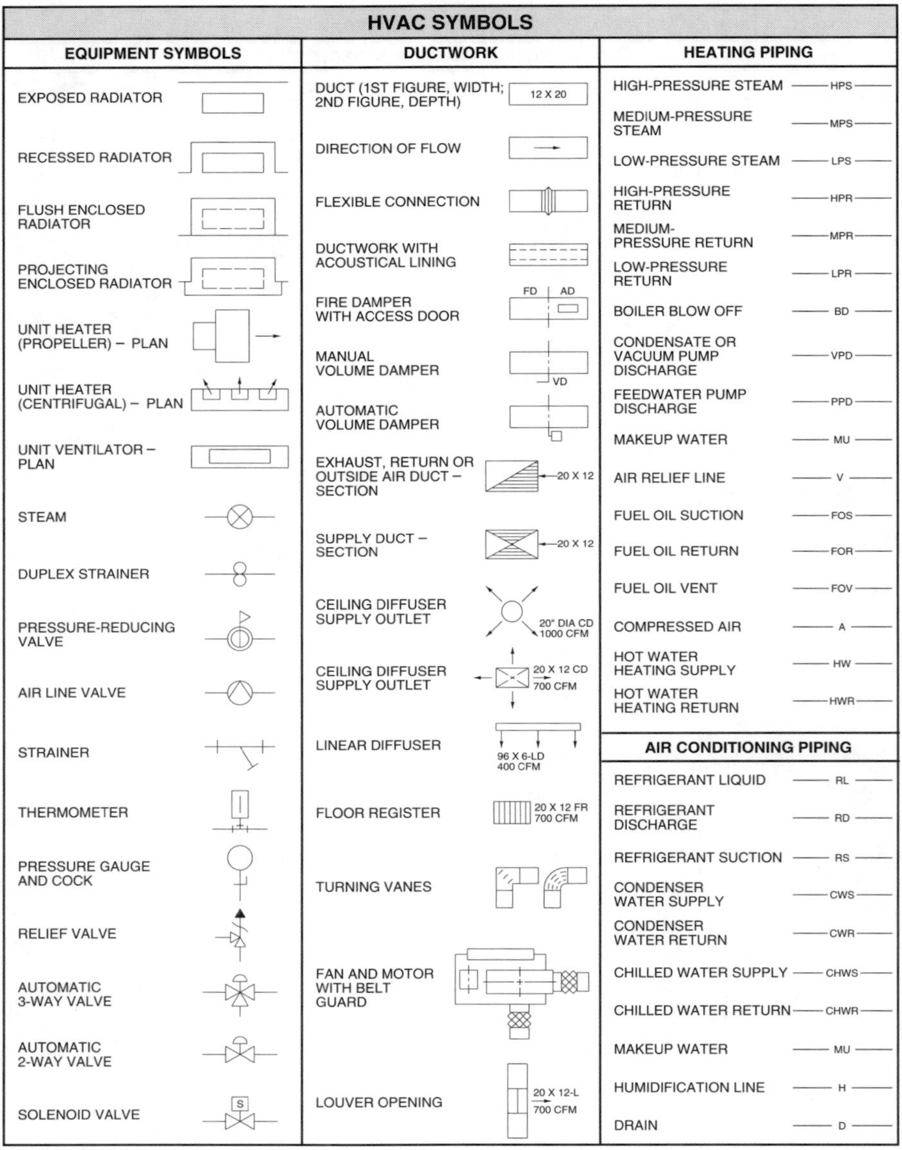

HVAC SYMBOLS

EQUIPMENT SYMBOLS

EXPOSED RADIATOR
RECESSED RADIATOR
FLUSH ENCLOSED RADIATOR
PROJECTING ENCLOSED RADIATOR
UNIT HEATER (PROPELLER) – PLAN
UNIT HEATER (CENTRIFUGAL) – PLAN
UNIT VENTILATOR – PLAN
STEAM
DUPLEX STRAINER
PRESSURE-REDUCING VALVE
AIR LINE VALVE
STRAINER
THERMOMETER
PRESSURE GAUGE AND COCK
RELIEF VALVE
AUTOMATIC 3-WAY VALVE
AUTOMATIC 2-WAY VALVE
SOLENOID VALVE

DUCTWORK

Symbol	Label
DUCT (1ST FIGURE, WIDTH; 2ND FIGURE, DEPTH)	12 X 20
DIRECTION OF FLOW	
FLEXIBLE CONNECTION	
DUCTWORK WITH ACOUSTICAL LINING	
FIRE DAMPER WITH ACCESS DOOR	FD AD
MANUAL VOLUME DAMPER	VD
AUTOMATIC VOLUME DAMPER	
EXHAUST, RETURN OR OUTSIDE AIR DUCT – SECTION	20 X 12
SUPPLY DUCT – SECTION	20 X 12
CEILING DIFFUSER SUPPLY OUTLET	20" DIA CD 1000 CFM
CEILING DIFFUSER SUPPLY OUTLET	20 X 12 CD 700 CFM
LINEAR DIFFUSER	96 X 6-LD 400 CFM
FLOOR REGISTER	20 X 12 FR 700 CFM
TURNING VANES	
FAN AND MOTOR WITH BELT GUARD	
LOUVER OPENING	20 X 12-L 700 CFM

HEATING PIPING

HIGH-PRESSURE STEAM	HPS
MEDIUM-PRESSURE STEAM	MPS
LOW-PRESSURE STEAM	LPS
HIGH-PRESSURE RETURN	HPR
MEDIUM-PRESSURE RETURN	MPR
LOW-PRESSURE RETURN	LPR
BOILER BLOW OFF	BD
CONDENSATE OR VACUUM PUMP DISCHARGE	VPD
FEEDWATER PUMP DISCHARGE	PPD
MAKEUP WATER	MU
AIR RELIEF LINE	V
FUEL OIL SUCTION	FOS
FUEL OIL RETURN	FOR
FUEL OIL VENT	FOV
COMPRESSED AIR	A
HOT WATER HEATING SUPPLY	HW
HOT WATER HEATING RETURN	HWR

AIR CONDITIONING PIPING

REFRIGERANT LIQUID	RL
REFRIGERANT DISCHARGE	RD
REFRIGERANT SUCTION	RS
CONDENSER WATER SUPPLY	CWS
CONDENSER WATER RETURN	CWR
CHILLED WATER SUPPLY	CHWS
CHILLED WATER RETURN	CHWR
MAKEUP WATER	MU
HUMIDIFICATION LINE	H
DRAIN	D

Glossary

acute angle: Angle with less than 90°.

acute triangle: A scalene triangle with each angle less than 90°.

addition: The process of uniting two or more numbers to make one number.

adhesive: A substance used to bond two surfaces together.

aerial platform: A lifting device for materials and workers.

altitude: Perpendicular dimension from the vertex to the base of a triangle.

angle: The intersection of two lines.

Arabic numerals: Numerals expressed by the ten digits 0, 1, 2, 3, 4, 5, 6, 7, 8, and 9.

arc: A portion of the circumference.

arc welder: A shielded metal arc welding (SMAW) machine in which the arc is shielded by the decomposition of the electrode covering.

area: The number of unit squares equal to the surface of an object.

backcut: A series of relief cuts made on the back surface to facilitate bending the piece.

backer board: Drywall installed in a suspended ceiling which serves as an attachment surface for acoustical tile.

base: The side upon which the triangle stands.

batten: A narrow strip of wood, metal, plastic, or drywall used to conceal an open joint.

bevel: An angle cut that extends from surface to surface of a piece of material.

carrier channels: The main supporting members of a suspended ceiling system to which furring channels are attached.

chalk line: A layout tool used to snap a straight line.

chamfer: An angled cut that extends from the surface to the edge of a piece of material.

chord: A line from circumference to circumference not through the centerpoint of a circle.

circle: A plane figure generated around a centerpoint.

circle cutter: A hand tool used to cut circles in thin wood or drywall sheets.

circumference: 3.1416 times the diameter of a circle.

claw hammer: A striking tool with a slightly curved head used to drive nails and a slotted claw used to pull nails.

common wall: An interior wall shared by two or more occupancies.

complementary angles: Two angles that equal 90°.

concentric circles: Circles with different diameters and the same centerpoint.

control-joint trim: A thin strip of perforated metal applied to relieve stress resulting from expansion and contraction in large ceiling and wall surfaces.

coreboard: A panel product consisting of a gypsum core encased with strong liner paper, forming a 1″ thick panel.

corner bead: A light-gauge, L-shaped piece of galvanized metal used to cover and protect the exposed outside corners of drywall.

corner skimming: Applying skim coats of joint topping compound to both surfaces of an inside corner joint.

cut-off saw: A light-duty portable electric saw used for cutting material to length.

decimal: A fraction with a denominator of 10, 100, 1000, etc.

decimal point: The period in a decimal number.

demountable partitions: A partition designed to be assembled, disassembled, and reassembled in another location with minimal damage to the partition components.

demountable partition system: A wall system which uses components designed to be disassembled and reused.

denominator: The part of a fraction that shows how many parts the whole number has been divided into.

diameter: The distance from circumference (outside) to circumference through the centerpoint of a circle.

dividend: The number to be divided.

division: The process of finding how many times one number contains the other number.

divisor: The number by which division is done.

double-layer drywall applications: Applications in which two layers of drywall are installed.

draftstop: The divider or partition in the attic of a structure which retards the spread of fire and smoke within the building.

drywall: Interior surfacing material applied to framing members using dry construction methods such as adhesive or mechanical fasteners.

drywall cutout tool: A hand-held electric cutout tool used to make irregular cuts and holes in panels, gypsum board, etc. for electrical boxes, duct openings, etc.

drywall hammer (drywall axe or hatchet): A striking tool with a serrated face used to drive fasteners into drywall and leave a dimple.

drywall lifter (kicker): A lifting device made of a short piece of metal that is tapered on one end and has a fulcrum on the bottom.

E

eccentric circles: Circles with different diameters and different centerpoints.

electric screwdriver (screwgun): A power tool used for driving various types of screws.

equation: A means of showing that two numbers or two groups of numbers are equal to the same amount.

equilateral triangle: A triangle that has three equal angles and three equal sides.

even numbers: Any numbers that can be divided by 2 an exact number of times.

exterior ceiling and soffit drywall: A moisture-resistant, gypsum-based sheet material with thicker face and back paper treated to resist mildew.

exterior insulation and finish systems (EIFS): Exterior panel systems composed of sheathing, insulation board, and a finish.

F

finished ceiling height: The height of the ceiling after the drywall sheets and any other covering have been applied.

fire protection rating: The length of time an assembly remains intact when exposed to fire.

fire taping: Drywall taping which is concealed behind paneling or above the finished ceiling to achieve the required fire rating.

flexible corner bead: A corner bead with flanges that may be cut to allow the corner bead to be bent to any curved radius.

formula: A mathematical equation which contains a fact, rule, or principle.

fraction: One part of a whole number.

fulcrum: The support about which a lever turns.

furring channels: Metal channels fastened to a structural surface to provide a base for fastening finish material.

G

gypsum board stripper: A hand-held cutting tool that cuts both sides of the panel at the same time.

gypsum shaftwall liner (coreboard): A gypsum-based sheet material with either tongue-and-groove or square edges.

gypsum sheathing: Exterior wallboard consisting of a water-repellent gypsum core with a water-repellent paper on face and back surfaces.

hacksaw: A metal-cutting handsaw with an adjustable steel frame for holding various lengths and types of blades.

hand tool: Any tool powered by a human.

hanger: A screw head that protrudes past the surface of the drywall face paper and prevents the drywall surface from being properly finished.

hanger wire: The wire that supports a suspended member. Hanger wire is supplied in rolls or bundles.

horizontal line: A line that is parallel to the horizon.

hypotenuse: The side of a right triangle opposite the right angle.

improper fraction: A fraction that has a numerator larger than its denominator.

inclined (slanted) line: A line that is neither horizontal nor vertical.

irregular plane figure: A plane figure that does not have equal angles and equal sides.

irregular polygon: A plane figure that contains unequal sides and unequal angles.

isosceles triangle: A triangle that contains two equal angles and two equal sides.

joint compound mixer (stomper): A long-handled hand tool used for mixing powder and premixed joint compound.

keyhole saw: A handsaw with thin, tapered, interchangeable blades used for cutting curves and inside holes.

laser level: A leveling device in which a concentrated beam of light is projected horizontally or vertically from the source and used as a reference for leveling or verifying horizontal or vertical alignment.

light trough: A recessed area in the ceiling which provides a concealed location for the installation of light fixtures used for indirect lighting.

locking "C" clamps: C-shaped clamps used to apply pressure to material.

millions period: The third period (1,000,000 through 999,999,999).

minuend: The number from which the subtraction is made.

mixed number: A combination of a whole number and a fraction.

moisture-resistant drywall: Drywall in which the gypsum core contains asphalt and other materials which can withstand moisture and face and back paper which is treated to resist rot and mildew.

multiplicand: The number which is multiplied.

multiplication: The process of adding one number as many times as there are units in the other number.

multiplier: The number by which multiplication is done.

N

nail pops: Blemishes caused when drywall nail heads force the finishing material past the surrounding surface and become exposed.

nail set: A small steel punch-like tool used to set finish nails below the surface of a trim member.

nail spotting: Covering the drywall screws or nails with several coats of joint topping compound.

numerator: Part of a fraction that shows the number of parts in the fraction.

obtuse angle: An angle that contains more than 90°.

obtuse triangle: A scalene triangle with one angle greater than 90°.

odd numbers: Any numbers that cannot be divided by 2 an exact number of times.

off angle: An angle greater or less than 90°.

½″ electric drill motor: A power tool used for drilling holes.

panel lifter: A roll-about lifting mechanism used for lifting and positioning drywall sheets to sidewalls and ceilings.

parallel lines: Lines that remain the same distance apart.

parallelogram: A four-sided plane figure with opposite sides parallel and equal.

period: A group of three digits separated from other periods by a comma.

perpendicular: A line or plane that makes a right angle with another (not necessarily horizontal) line or plane.

Phillips head screwdriver: A hand tool with a head designed to fit into a cross-slotted screw head for turning.

pipe-hole cutter: A hand tool used to cut holes in drywall sheets that are too small to cut with a circle cutter.

plane figure: A flat figure.

plumb: An exact verticality (determined by a plumb bob and line) with Earth's gravity.

plumb bob: A layout tool used to establish a vertical line.

pneumatic tool: A tool powered by compressed air.

polarity: The particular state of an object, either positive or negative, which refers to the two electrical poles, north and south.

pole sander: A sanding tool with sandpaper attached to a swivel head on a pole.

polygon: A many-sided plane figure.

powder-actuated tool: A device that drives fasteners by means of an explosive charge.

power tool: Any tool powered by a source other than humans.

prefinished drywall: Dywall which is covered with vinyl, fabric, or other materials.

prime numbers: Numbers that can be divided an exact number of times only by themselves and the number 1.

product: The result of multiplication.

proper fraction: A fraction that has a denominator larger than its numerator.

Pythagorean Theorem: The square of the hypotenuse of a right triangle is equal to the sum of the squares of the other two sides.

quadrant: One-fourth of a circle.

quadrilateral: A four-sided polygon with four interior angles.

quotient: The result of division.

radius: One-half the length of the diameter of a circle. The distance from the centerpoint to the outer edge of a circle.

rasp: A coarse file used to shape wood, drywall, and other soft materials.

rectangle: A quadrilateral with opposite sides equal and four 90° angles.

regular plane figure: A plane figure that contains equal angles and equal sides.

regular polygon: A polygon that contains equal sides and equal angles.

remainder: In subtraction, the difference between the minuend and the subtrahend. In division, the part of the quotient left over whenever the quotient is not a whole number.

resilient channel: A preformed metal channel which maintains an air space between the framing member and the drywall sheet.

reveal: A solid channel-shaped strip of metal with perforated surfaces on both sides.

reveal trim: A drywall joint that is open to view.

rhomboid: A quadrilateral with opposite sides equal and no 90° angles.

rhombus: A quadrilateral with all sides equal and no 90° angles.

right angle: An angle that contains 90°.

right triangle: A triangle that contains one 90° angle and no equal sides.

Roman numerals: Numerals expressed by the letters I, V, X, L, C, D, and M.

rough ceiling height: The height of the ceiling before the drywall sheets and any other covering have been applied.

scalene triangle: A triangle that has no equal angles or equal sides.

secant: A straight line touching the circumference of a circle at two points.

sector: A pie-shaped piece of a circle.

segment: The portion of a circle set off by a chord.

semicircle: One-half of a circle.

shaftwall: A rated enclosure which encloses elevators, air ducts, plumbing pipes, electrical wires, or other items which pass through the floors of a high-rise building.

Sheetrock®: A brand of gypsum panel developed by the United States Gypsum Company for interior wall and ceiling surfaces.

skim coat: A thin coat of joint topping compound.

snips: A scissor-like hand tool used for cutting light-gauge metal and other materials.

soffit: An area of the ceiling that is constructed at a finished height lower than the main ceiling.

spirit level: A layout tool used to establish and check vertical and horizontal lines.

square: A quadrilateral with all sides equal and four 90° angles.

square foot: A plane area that contains 144 sq in. (12″ × 12″ = 144 sq in.).

square inch: A plane area that measures 1″ × 1″ or its equivalent.

stapler: A pneumatic tool which drives staples into drywall and wood.

steel square: An L-shaped layout tool with or without graduations.

straight angle: An angle that contains 180°.

straight line: The shortest distance between two points.

strut: A member fixed between two other members to maintain a specified distance.

subtraction: The process of taking one number away from another number.

subtrahend: The number which is subtracted.

sum: The result obtained from adding two or more numbers.

supplementary angles: Two angles that equal 180°.

tangent: A straight line touching the circumference of a circle at only one point.

tape measure (steel tape): A hand-held measuring device with a retractable, graduated steel blade.

taping: Applying joint tape over joint compound to all drywall joints.

taping knife: A hand tool used to wipe down the tape after it has been applied.

taping (mud) pan: A joint compound carrier for hand finishing.

texturing: Applying an uneven surface to the drywall.

thousands period: The second period (1,000 through 999,999).

T-nailer: A pneumatic tool which drives hardened T-nails into concrete or masonry surfaces.

tool: Any device used to produce work.

trapezium: A quadrilateral with no sides parallel.

trapezoid: A quadrilateral with two sides parallel.

triangle: A three-sided polygon with three interior angles.

T-square (drywall T-square): A T-shaped layout tool with or without graduations.

units period: The first period (000 through 999).

utility knife: A cutting tool with a short blade protruding from the handle.

vertical: A line in a straight upward position.

vertical line: A line that is perpendicular to the horizon.

W

whole numbers: All numbers that have no fractional or decimal parts.

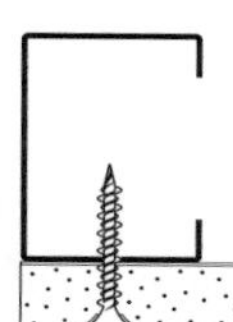

Index

acoustical ceiling drywall grid, 52
acute angle, 16 – *17*
addition, 4 – *5*
  decimals, 15
  fractions, 8 – 10
adhesive installation, 111 – 114
adhesives, 33 – 36
all-purpose joint compound, 55
altitude, 18
angles, 16 – 17
  acute, 16 – *17*
  complementary, 16 – 17
  obtuse, 16 – *17*
  right, 16 – *17*
  straight, 16 – *17*
  supplementary, 16 – *17*
Arabic numerals, *2*
arc, *17*

backer board, 31
base, 18
batten, 33
Bazooka®, 181 – 183
bends, 161 – *163*
bevel, 172
bracing, *46*, 48 – *49*
building permit, *104*

carrier channels, 51
centerpoint, *17*
chalk line, *66*
chamfer, *172*
chord, *17*
circle cutter, *73* – 74
circles, *17*
  area, 23
  concentric, *17*
  eccentric, *17*
  full, *17*
  semicircles, *17*
circumference, 17
claw hammer, *68* – 69
coating metal trims, 187 – *188*
commercial ceiling textures, 190
common fractions, 7 – 13
  changing decimals to, 15
complementary angle, 16 – *17*
contact adhesives, 34 – 35
control-joint trim, *38*, *116*
coreboard, 31
corner bead, 36 – *37*
corner skimming, *187*
corridors, 147 – *148*
curved line, *16*
curves, *163* – *164*
cut-off saw, *83*

decimal point, 14
decimals, *14* – 16
  adding, 15
  dividing, 15
  multiplying, 15
  subtracting, 15
demountable partitions, 32 – *33*, 137 – *138*
  installation, 137 – *138*
denominator, 7
diameter, 17
dividend, 7
division, *7*
  decimals, 15
  fractions, 12 – 13
divisor, 7
draftstops, 152 – *153*
drywall, 29 – 33
  core composition, 30
  cutout tool, *82*
  edges, 31 – *32*
  hammer, *69* – 70
  handling, 101 – 103
  installation, 103 – 117
  lifter, 75 – *76*
  manufacture, 29 – 30
  multiple layers, 113 – 114
  nails, *40* – 42, 106 – 107
  screws, *43* – 44

sheet selection, 105 – 106
sizes, 31
staples, *44*
storage, *102*
taping and finishing materials, 54 – 56
thicknesses, *31*
trims, 36 – 40

electrical cords, 86 – 87
electric screwdriver (screwgun), *80 – 81*
elevator shafts, 149 – *150*
equation, 20
even numbers, 1
exterior ceiling and soffit drywall, 169 – *170*
taping, 184 – 185
exterior fire-control assemblies, *154* – 156

face and back paper, 30
F-corner trim, *39*
fiberglass joint tape, 54 – *55*
finished ceiling height, 128
finishing, 185 – 189
fire protection rating, 147
fire-resistant chases, *151 – 152*
fire taping, 184
formulas, 20 – 24
area, *22* – 24
circumference, 22 – 23
electrical, 20
Pythagorean theorem, *24*
fractions, 7 – 13
adding, 8 – *9*
changing decimals to, 15
dividing, 12 – *13*
multiplying, *11* – 12
subtracting, *10* – 11
furring channels, 51 – *52*
ceilings, *127 – 131*

garage walls, 150
Greek letters, *21*
gypsum board stripper, *74*
gypsum panels, 29 – 44
gypsum shaftwall liner, 168 – *169*
gypsum sheathing, 31, 170 – *172*

hacksaw, 72 – *73*
hand tools, 61 – 79
hanger, *110*
hanger wire, 50 – *51*, 130 – 131
heavy-gauge metal framing materials, *45 – 46*
heavy-gauge metal studs, 45 – 46
heavy-gauge metal stud walls, 124 – 127
framing inspections, 126 – 127
layout and track installation, 124 – *125*
stud installation, 125 – *126*
horizontal line, 16

improper fraction, 8
adding, 9
inclined (slanted) line, *16*
interior fire-control assemblies, 146 – 154
installation, 153 – 154

joint compound, *35* – 36, 55
joint compound mixer, *78*
joint topping compound, 55 – 56
joist-framed ceiling, 52 – 53
J-trim, *37*, *115*

keyhole saw, *72*

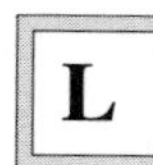

laser level, 85 – *86*
light-gauge metal framing materials, *47* – 50
light-gauge metal studs, 48
light-gauge metal stud walls, 121 – 124
framing inspections, 124
stud installation, 122 – *123*
light troughs, 131 – *133*
lines, *16*
curved, *16*
horizontal, 16
inclined, *16*

parallel, *16*
straight, 16
vertical, 16
locking "C" clamps, *75*
lowest common denominator, 8
L-trim, *37*, 114 – *115*

materials, 29 – 60
measurement systems, 3
mechanical taping tools, 78 – *79*
metal ceiling framing materials, 50 – 54
metal framing members, 45 – 50
cutting, *48*
metal framing safety precautions, 53 – 54
metal stud track, 46
metal stud walls, 121 – 127
metric system, 3 – *4*
millions period, 2
minuend, 5
mixed number, 7
adding, 9
dividing, 12 – 13
multiplying, 12
subtracting, 11
moisture-resistant drywall, 166 – *168*
taping, 185
multiplicand, 6
multiplication, *6*
decimals, 15
fractions, *11* – 12
multiplier, 6

nail-on installation, 105 – *107*
nail pops, 40 – 41
nails, *40* – 42, 106 – 107
cement-coated cooler, 42
chemically-etched, 41 – *42*
color pins, 42
nail spotting, *186* – 187
90° angles, 164 – *165*
numerator, 7

obtuse angle, 16 – *17*
odd numbers, 1
off angles, 165 – *166*
½″ electric drill motor, *82*

panel adhesives, 33 – *34*, 111 – 113
panel lifter, *76*
paper joint tape, *54*
parallel lines, *16*
parallelogram, *19*
period, 2
perpendicular, 16
personal protective equipment, *96*
Phillips head screwdriver, 74 – *75*
pipe-hole cutter, *74*
plane figures, 16 – 20
irregular, 16
regular, 16
planning and inspection checklist, *104*
plumb, 16
plumb bob, 66 – *67*
pneumatic tools, 92 – 93
safety, 95
pole sander, *78*
polygons, 19 – *20*
irregular, 19
regular, 19
portable arc welder, 83 – 85
powder-actuated tools, 87 – 92
cartridges, 89 – 91
direct-acting, *88* – 89
fastener length, 97
indirect-acting, *88* – 89
safety, 91 – 92, 95
power tools, 79 – 87
prefinished drywall, *172* – *174*
prefinished drywall trim, 38 – 40, *117*
prime numbers, 1
product, 6
proper fraction, 7
adding, 8
Pythagorean theorem, *24*

quadrant, *17*
quadrilaterals, 18 – *19*
quick-setting joint and topping compound, 56
quotient, 7

radius, *17*
rasp, *71*

rectangle, *19*
  area, 23
remainder, 5, 7
residential ceiling textures, 189 – 190
reveal trim, *38*, 116 – *117*
rhomboid, *19*
rhombus, *19*
rolling corners, *184*
Roman numerals, 2 – *3*
rough ceiling height, 128

safety, 94 – 96
sanding, *188*
sanding pole, 188
scaffold and ladder safety, 95 – 96, 155 – 156
screw-on installation, 108 – 111
  sequence, 109 – 110
screws, *43* – 44
secant, *17*
sector, *17*
semicircle, *17*
shaftwall, 49 – *50*, 134 – *136*
  framing, 49
  installation, 135 – *136*
Sheetrock®, 29
skim coat, 185 – 186, 188 – 189
  no. 1, 185
  no. 3, 186
  no. 2, 186
snips, 71 – *72*
soffits, 131 – 133
sound-control assemblies, 143 – *146*
  installation, *146*
special applications, 166 – 174
spirit level, 67 – *68*
square, *19*
  area, 23
stairways, 148
stapler, 92 – *93*
staples, *44*
steel square, 63 – *64*
straight angle, 16 – *17*
straight line, 16
structural steel beams and columns, *150*
strut, 34
subtraction, *5* – 6
  decimals, 15
  fractions, *10* – 11
subtrahend, 5
supplementary angle, 16 – *17*
suspended ceiling framing materials, 50 – 52

tangent, *17*
tape measure, 61 – 63
taping, 179 – 185
  inspection, 179 – *180*
  installation, 181 – 185
  preparation, *180* – *181*
taping and finishing materials, 54 – 56
taping knife, *77*
taping (mud) pan, 76 – *77*
texturing, 189 – 190
thousands period, 2
T-nailer, *93*
tools, 61 – 93
trade math, 1 – 28
trapezium, *19*
trapezoid, *19*
triangles, 18
  acute, *18*
  area, 23 – 24
  equilateral, *18*
  isosceles, *18*
  obtuse, *18*
  right, *18*
  scalene, *18*
trim installation, 114 – 117
T-square, 64 – 66, *94*

units period, 2
utility holes, 45, 47
utility knife, *70* – 71
U-trim, *37*, *115*

vertical, 16
vertical line, 16

whole numbers, 1
wiping down tape, *183*